CELL BIOLOGY RESEARCH PROGRESS

LIPID RAFTS

PROPERTIES, CONTROVERSIES AND ROLES IN SIGNAL TRANSDUCTION

CELL BIOLOGY RESEARCH PROGRESS

Additional books in this series can be found on Nova's website under the Series tab.

Additional e-books in this series can be found on Nova's website under the e-book tab.

CELL BIOLOGY RESEARCH PROGRESS

LIPID RAFTS

PROPERTIES, CONTROVERSIES AND ROLES IN SIGNAL TRANSDUCTION

DAN SILLENCE
EDITOR

NOVA BIOMEDICAL

New York

For permission to use material from this book please contact us:
Telephone 631-231-7269; Fax 631-231-8175
Web Site: http://www.novapublishers.com

Library of Congress Cataloging-in-Publication Data

ISBN: 978-1-62948-961-2

Library of Congress Control Number: 2013956784

Published by Nova Science Publishers, Inc. † New York

Contents

Preface

Greater than 1000 different lipids occur in cells and there is a quest to understand this complexity. Some lipids are essential as knockout mice fail to develop. In the case of many lipids this is due to their structural importance in plasma membrane permeability (e.g. cholesterol and sphingomyelin). In a few instances essential lipids take a direct part in a signaling cascade (examples include phosphoinositides, phosphatidylserine and diacylglycerol). However, sphingolipids, cholesterol and glycosylphosphatidylinositol(GPI)-anchored proteins are thought to be enriched in self organizing platforms on the surface of cells (lipid rafts). It would be anticipated that molecules which need to be in close proximity to signaling and transport machinery may fail if dispersed randomly. These platforms can facilitate the clustering of cell surface receptors for efficient signaling and uptake into the cell. Lipid rafts are also be expected to span the bilayer transducing signals to the opposite side.

The potential importance of lipid rafts started with an extreme example of a lipid domain, the apical surface of polarized epithelial cells. The apical surface has a distinct lipid and protein composition in comparison to the opposite basolateral membrane and thus can be looked at as one big lipid domain. The lipid raft hypothesis explained how lipids synthesized inside the cell could be sorted and concentrated en route to the apical membrane. Lipid rafts were proposed to be self-organising platform that once synthesized could be transported to the apical surface.

Lipid rafts were initially purified by detergent solubilisation of epithelial cells through purification of structures that had a similar lipid composition to the apical surface (cholesterol:sphingolpid:glycerolipid 1:1:1). This technique was then widely used in other cells to purify lipid rafts. Although a wide variety of signaling proteins and lipids have also been found to be enriched detergent resistant structures difficulties in visualizing these structures in intact cells has led to considerable controversy as it was not clear whether the structures purified were biologically relevant pre-existing structures. Another structure that is believed to be related to lipid rafts are caveolae which are small invaginations on the surface of cells.

It is now clear that a variety of high resolution microscopic and mass spectrometric techniques are able to detect domains on the surface of cells. Although it is still not clear exactly how they are formed, this book addresses the functions of these domains in normal cell processes (e.g. signal transduction) and how they are disrupted in disease.

In the following book glycosphingolipid candidates important in the formation of lipid rafts are identified by gene knockout technology (Koichi, Nagoya). The biophysical basis of lipid rafts are then described (Silva, Lisboa) then the role of caveolae in K^+ channel regulation (Dart, Liverpool) and energy homeostasis (Hundal, Dundee). This is followed by chapters in the role of lipid rafts in immune cell function and dysfunction (Jury, London) as well as mast cell signaling (Mazucato, Sau Palou). The role of lipid rafts in neurodegenerative diseases such as Alzheimer's and Parkinson's are then described (Bate, Hertfordshire; UK; Marin, Tenerife).

In: Lipid Rafts
Editor: Dan Sillence
ISBN: 978-1-62948-961-2

Chapter I

The Role of Glycosphingolipids in Lipid Rafts: Lessons from Knockout Mice

Koichi Furukawa[1*], Yusuke Ohmi[1], Yuji Kondo[1], Yuki Ohkawa[1], Noboru Hashimoto[1], Orie Tajima[2] and Keiko Furukawa[*1,2]
[1]Department of Biochemistry II,
Nagoya University Graduate School of Medicine, Showa-ku, Nagoya
[2]Department of Biomedical Sciences,
College of Life and Health Sciences, Chubu University, Kasugai, Japan

Abstract

Lipid rafts have been analyzed mainly by cultured cells, and by biochemical analysis using cell extracts with various detergents. Recent progress in the technology for gene-engineering of glycosyltransferase genes and for single molecule imaging with high spatio-temporal resolution is altering concepts of lipid rafts. In particular, gene targeting of glycosyltransferases has enabled us to investigate *in vivo* function of glycosphingolipids (GSLs), in order to confirm proposed functions based on *in vitro* studies. Results obtained using gene knockout (KO) mice have

[*] Corresponding author: Koichi Furukawa: Tel: 81-52-744-2070; Fax: 81-52-744-2069; E-mail: koichi@med.nagoya-u.ac.jp.

revealed novel functions. The majority of abnormal phenotypes may be attributed to aberrant architecture and function of lipid rafts due to deficiency of GSLs. However, abnormal phenotypes exhibited by KO mice were generally milder than expected, suggesting that remaining GSLs can compensate for function of lost structures, functional redundancy. However, some functions cannot be replaced by remaining GSLs in the KO mice. Thus, there may be two kinds of function of GSLs in lipid rafts. One is those which can be compensated, not specific, and no specific ligand molecules have been identified. Total amounts of GSLs should be important in this case. The other are those which can not be compensated by other structures, are specific, and need specific ligand molecules to exert their role. Combination of various KO mice should be effective for the clarification of functional redundancy among GSLs. Identification of physiological ligands to individual GSLs is also crucial to investigate implication of each specific function.

Keywords: Glycosphingolipids, lipid rafts, microdomains, gangliosides, knockout

Introduction

Among biomolecules glycosphingolipids (GSLs) are quite unique in many aspects. They are amphipathic molecules consisting of hydrophilic carbohydrate moieties and hydrophobic lipid moieties such as ceramide or glycerol [1]. The carbohydrate moiety is polymorphic with mainly 5 major core structures and lineages. Namely, ganglioside-series, globo-series, lacto/neolacto-series, and asialo-series glycolipids, all of which are synthesized via glucosylceramide (Glc-Cer). Besides these GSLs, gala-series GSLs generated via galactosylceramide (Gal-Cer) are also present. Addition of galactose with β1,4-linkage results in the synthesis of lactosylceramide (Lac-Cer), that is a common precursor of almost all GSLs. On the other hand, Gal-Cer is an initiating structure of the other group of GSLs, the number and variety of GSLs derived from Gal-Cer are very limited. In addition to the polymorphism in the carbohydrate moiety of GSLs, heterogeneity in the ceramide portion has been known for long time. Recently, implication of the ceramide portion has been paid attention to not only as structural features among species and/or tissue differentiation, but also as functional roles in the significance of whole GSL molecules.

Although a number of biological functions for individual GSL structures have been reported until 1990, they were fragmental and based on restricted experimental conditions. Furthermore, aspects of structural features had been focused on rather than biological functions. Since cDNAs of glycosyltransferase responsible for the synthesis of GSLs was first isolated [2], functional analysis of GSLs has dramatically progressed. Indeed, a number of glycosyltransferase cDNAs have become available by the last decade of 20th century [3]. The synthetic pathway and enzymes for GSLs synthesis are shown in Fig. 1. Subsequently, since the middle of the 1990s, studies of knockout (KO) animals, in which glycosyltransferase or catalytic enzyme genes were genetically disrupted, have been reported [4]. Findings on abnormal phenotypes of various glycosyltransferase gene KO mice have provided us with novel information about ever-hidden roles of GSLs. Simultaneously, KO mice of some glycosyltransferase genes sometimes revealed no or much milder phenotypes than expected [5].

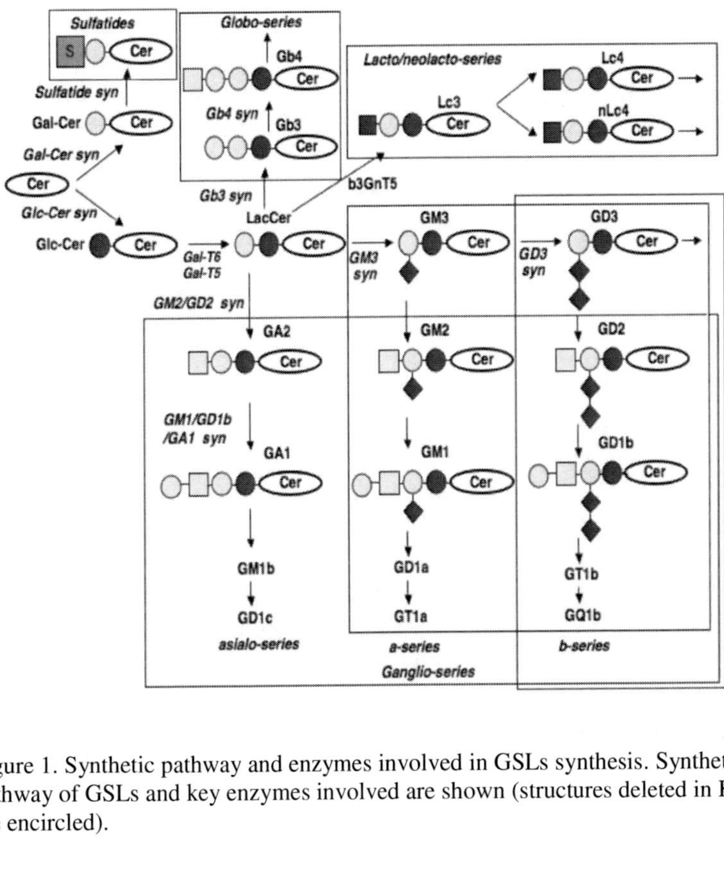

Figure 1. Synthetic pathway and enzymes involved in GSLs synthesis. Synthetic pathway of GSLs and key enzymes involved are shown (structures deleted in KO mice are encircled).

In this review, we have tried to briefly summarize the main findings of individual KO mice of GSL synthase genes, and also to clarify what have been demonstrated for the structures and functions of lipid rafts in association with their abnormal phenotypes.

1. Gangliosides Modulate Signals Transmitted by Neurotrophic Factors

Gangliosides are highly expressed and have been considered to be involved in the development, differentiation and function of nervous systems [6]. A number of studies have been reported to suggest that gangliosides play roles as neurotrophic factors using cultured neural cell lines [7]. Our study using a rat pheochromocytoma cell line PC12 demonstrated that gangliosides regulate proliferation [8] and differentiation [9] of these cells by modulating intracellular localization (inside/outside of lipid rafts) of NGF receptor TrkA and other signaling molecules, leading to the dramatic alteration in the response of cells to NGF. These results showed roles of gangliosides in the regulation of signals that are transduced via lipid rafts.

2. Important Roles of GSLs in the Regulation of Multi-Cellular Organisms

Although GSLs were not essential in the growth and survival of mammalian cells as shown by a Glc-Cer synthase-deficient cell line [10], genetic desruption of Glc-Cer revealed that GSLs were needed in the early stage of development, suggesting their roles in the regulation of multi-cellular bio-organism [11]. The null mutant mice of the GlcCer synthase showed embryonic lethality by day E7.5. These results suggest that correct carbohydrate structures of GSLs are essential for normal embryonic development and for the differentiation of various tissues. Conditional knockout (KO) mice of Glc-Cer synthase, in which this gene was disrupted after birth, showed dysfunction of cerebellum and peripheral nerves associated with structural defects [12]. They died within 3 weeks after birth, suggesting that GSLs are indispensable for the maintenance of life.

3. Essential Roles of Gal-Cer-Derived GSLs in Some Aspects of CNS Function

Galactosylceramide synthase (ceramide galactosyltransferase, CGT)[13,14] is responsible for the synthesis of galactosylceramide from ceramide and its sulfated derivative, sulfatide. Both of them are enriched in myelin. The null mutant mice of CGT could, however, generate myelin containing glucocerebroside [15]. Monogalactosyl- alkylacylglycerol, a precursor of seminolipid, is also defective in addition to Gal-Cer and sulfatide in KO mice. Severe generalized tremor, mild ataxia and conduction deficits were observed in this KO mouse [15]. Moreover, the mutant mice developed hindlimb paralysis and extensive vacuolation of the spinal cord with aging, suggesting that GalCer and/or sulfatide play important roles in myelin. This mutant also showed male sterility [16], suggesting that the products of this enzyme are essential for the spermatocyte passage into the first meiotic division.

The sulfatide synthase is a single gene for the synthesis of sulfatide and seminolipids [17], and they were completely deleted in KO mice of sulfatide synthase [18]. Sulfatides are expressed mainly in oligodendrocytes in CNS, and in Schwann cells in the peripheral nerves. As expected, KO of sulfatide synthase resulted in similar phenotypes to GalCer synthase disruption, i.e. neurological abnormalities and disrupted spermatogenesis [18]. However, these dysfunctions were less severe and appeared to occur later than those in CGT null mice, suggesting the unique roles of GalCer itself.

4. Knockout Mice of GM3 Synthase

GM3 synthase is an essntial enzyme for the synthesis of GM3 from lactosylceramide, and therefore, for the synthesis of all ganglio-series gangliosides. KO mice of GM3 synthase exhibited no marked abnormal phenotypes except increased sensitivity to insulin [19]. This phenotype appears to be different from the features of patients with GM3 synthase deficiency [20]. GM3 synthase-deficient patients in two families exhibited "infantile-onset symptomatic epilepsy" with growth and mental retardation and ophthalmologic deficits. This should be mainly because the involvement of asialo-series gangliosides is more significant in mice, and they can compensate fairly broad range of functions of the lost structures. Metabolic features of GSLs in these situations are shown in Fig. 2. The KO mice, however, showed increased sensitivity to insulin due to the enhanced phosphorylation of insulin receptor in skeletal muscle. These results well corresponded with roles of GM3 in the regulation of insulin/insulin receptor-mediated signals reported later [21]. Possibly, GM3 binds juxta-membrane region of insulin receptor in lipid rafts.

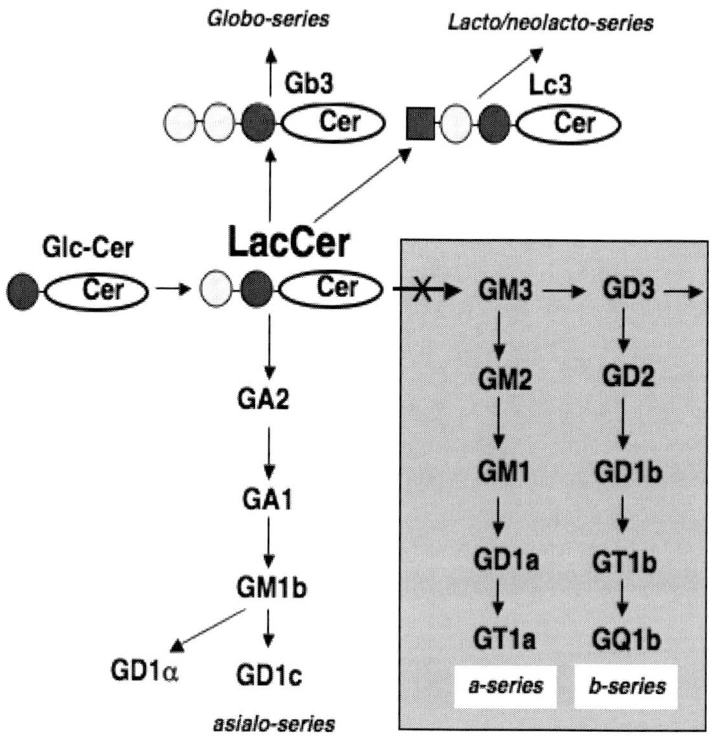

Figure 2. Metabolic features of GSLs in GM3 synthase KO mice and GM3 synthase deficient patients. All a-series and b-series gangliosides were deleted as indicated by dark squares. GM3 synthase KO resulted in the marked increase of asialo-series gangliosides in brain tissues. On the other hand, patients with GM3 synthase deficiency showed increase in globo- and lacto/neolacto-series GSLs probably due to the difference in the substrate specificity of GM2/GD2 synthase for LacCer.

5. Knockout Mice of Lactosylceramide Synthase

Genetic disruption of lactosylceramide synthase (β1,4Gal-T6) resulted in no definite abnormal phenotypes [22]. On the other hand, KO mice of β1,4Gal-T5 showed severe defects in the development of fetus as GlcCer synthase KO mice showed [11]. Accordingly, it has been shown that β1,4Gal-T5 might be main enzyme for the LacCer synthesis [23, 24], although β1,4Gal-T6 may be responsible for LacCer synthesis in particular environments [22]. It remains to be clarified why β1,4Gal-T6 was isolated as a representative LacCer synthase in the study of LacCer synthase purification and molecular cloning using rat brain [25].

6. Disruption of Ganglioside Synthases Resulted in the Milder Phenotypes than Expected

The generation and analysis of KO mice lacking the GM2/GD2 synthase gene was first reported by our group in 1996, and demonstrated that mutant mice lacked all complex gangliosides, but formed almost normal architectures of the central nervous system (CNS) [26]. There were no clear differences from the wild type in brain morphology, myelination and behaviors except for reduced nerve conductivity in the mutant mice. Male infertility due to aspermatogenesis appeared to be the most serious phenotype. This finding led to the identification of a novel function of complex gangliosides in testosterone transport [27]. With long-term observation of KO mice, neuronal degeneration was observed to occur gradually [28]. These degenerative disorders progressed with aging, and were mainly detected in the peripheral nerves and in dorsal horn of the spinal cord [28] Neurodegeneration in this mutant mice is summarized in Fig. 3. Other groups generated the same KO mice and reported Wallelian degeneration and abnormal neurological function in a motor neuron-dominant manner [29, 30]. In turn, our KO mice showed neurodegeneration predominantly in the sensory nerves. Morphological changes in synaptic vesicles and dendrites in the central terminals, and in glia process suggested that compensatory modification of the nerve tissues occurred after nerve degeneration because of the lack of complex gangliosides

(Fig. 3). Furthermore, regeneration of resected hypoglossal nerves was markedly repressed [31]. These results indicated that complex gangliosides are essential in the maintenance and repair of nerve tissues but not in morphogenesis. Aberrant Ca^{2+}-regulating properties in the cerebellar neurons were also found in this KO mice [32] and may correspond with the neurological disorders described above.

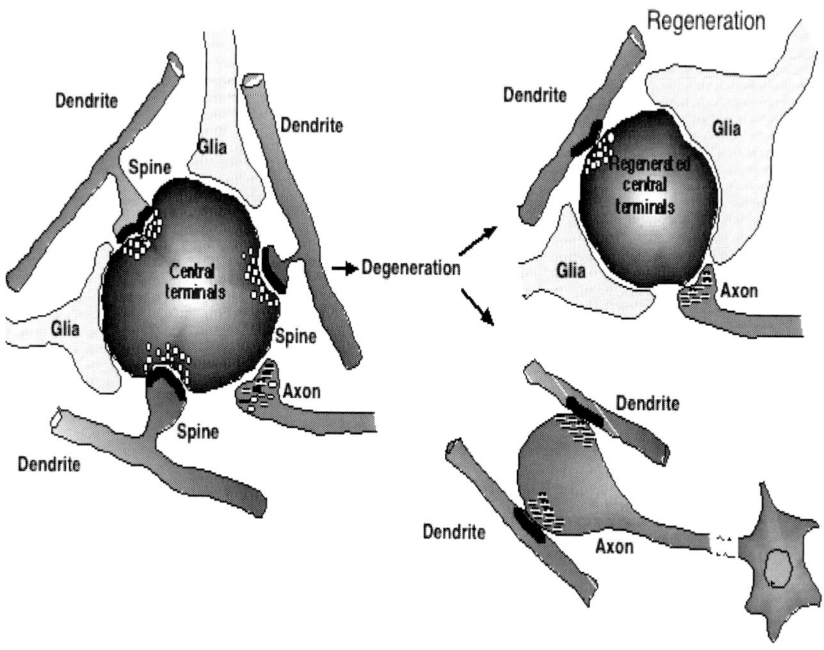

Figure 3. Neurodegeneration and regeneration in GM2/GD2 synthase KO. Degeneration of the central terminals occurred, leading to regeneration that modified the synapses, i.e. vesicles became flat from spherical shape and contact became dendrite-mode from spine-mode.

7. Knockout Mice of GD3 Synthase

GD3 synthase is a key enzyme for the synthesis of b-series and c-series gangliosides as shown in Fig. 1. Therefore, genetic disruption of this gene resulted in the loss of all b-series and c-series gangliosides [33, 34]. They showed almost no apparent abnormalities either in morphology or behavior.

Since GD3 was reported to mediate the apoptotic signals from Fas–Fas ligands [35], we examined the sensitivity of thymocytes from these KO mice to apoptosis induced by anti-Fas antibody, showing no lowered apoptosis induction. Although no morphological abnormalities were detected in nervous tissues in KO mice, apparently reduced regenerative activity was found in the hypoglossal nerve resection system [34], indicating crucial roles of b-series gangliosides in nerve regeneration. These results were in good agreement with the results of the regeneration experiments of the rat hypoglossal nerve system, in which b-series gangliosides showed the best effects on the regeneration of cleaved hypoglossal nerves [36].

8. Double (D) KO Mice of GM2/GD2 Synthase and GD3 Synthase Show Severe Phenotypes

As described above, both GM2/GD2 synthase KO and GD3 synthase DKO mice showed milder phenotypes than expected, probably due to the compensation by the remaining GSLs in the individual KO mice. As summarized in Fig. 4, profiles of GSL species in the individual KO mice revealed increases in the precursors of the lost GSLs, e.g. markedly increased GD3 and GM3 were found in brain tissues of GM2/GD2 synthase KO mice. When enzyme genes located in the initiating point for the synthesis of some lineage were disrupted, GSLs in other lineages increased, e.g. asialo-series GSLs were markedly increased in GM3 synthase KO mice. These results suggested two important possibilities. One is that whole amount of GSLs is strictly defined and total sum of GSLs is normalised. The other one is that functions of some GSLs can be replaced by other GSLs. In order to evaluate role of each GSL structure, we have tried to generate complex KO mice by mating different lines of KO mice, in which remaining GSLs are reduced as much as possible (Fig. 4).

In DKO mice lacking both GD3 synthase and GM2/GD2 synthase genes by mating these two mutants [37], no obvious changes were detected at birth, while they had only GM3 among all gangliosides. However, they gradually died of unknown reason after 12 weeks of birth. They also exhibited refractory skin lesions at face and neck. Repeated scratching action due to the reduced sensitivity to mechanical pain seemed to trigger the skin legions. Neurodegenerative changes in both peripheral nerves and CNS in DKO mice at a young age might explain the dysfunction in the sensory system (37). Audiogenic seizures and consequent sudden death by noise were observed in the same kind of DKO mice [38]. The differences in the resulting phenotypes in these DKO lines may be due to the different genetic backgrounds of the ES lines used for gene targeting, i.e. (C57BL/6xCBA) F1 versus 129. The phenotypic analyses of these mutant mice demonstrated that GM3 alone enables mice to undergo almost normal neurogenesis, birth and development up to a certain point of life. However, the correct composition of gangliosides is needed for the maintenance of intact morphology and function, and also for the repair of lesions in nerve tissues. Subsequent analysis to clarify the mechanisms for the neurodegeneration in DKO mice revealed that integrity in the architecture and function of nervous tissues can be maintained by complete lipid rafts based on correct composition of GSLs [39, 40].

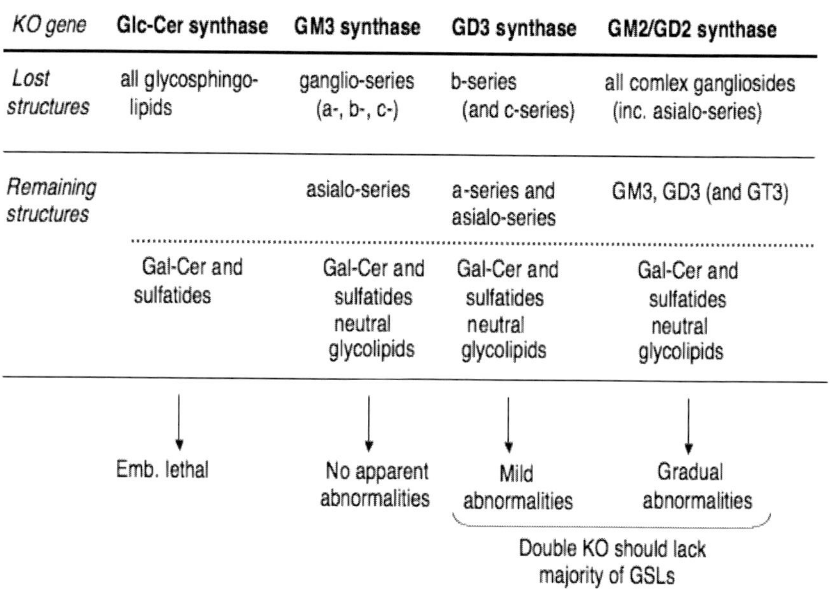

Figure 4. Lost structures and remaining structures in various KO mice. Remaining GSL species in the individual KO mice should be important to compensate for the function of lost structures in the individual KO mice.

Complex KO mice of GM3 synthase and GM2/GD2 synthase genes were also generated [41] in order to cancel the effects of asialo-series gangliosides, which remained in the GM3 synthase KO mice. The DKO mice showed severe nerve degeneration, leading to death. These results indicated that GM3 itself plays essential roles in the CNS.

9. Roles of Carbohydrate Structures in GSLs in the Regulation of Lipid Rafts

In order to overcome functional redundancy in the ganglioside species, DKO of two major glycosyltransferase, i.e. GD3 synthase and GM2/GD2 synthase genes were generated and analyzed as described above. They showed much more severe neurodegeneration than those detected in the single KO mice. In the analysis of mechanisms for the neurodegeneration, novel roles of gangliosides in the regulation of complement system were found [37]. Furthermore, roles of gangliosides in the formation and maintenance of membrane microdomains in brain tissues to control complement activation, suppression of inflammatory reaction and neurodegeneration were demonstrated [42]. Disruption of glycolipid-enriched microdomain (GEM)/rafts was compared among several mutant mice of ganglioside synthases, and up-regulation of complement-related genes corresponding with proliferation of astrocytes and infiltration of microglia were found with gradual severity depending on the defects in ganglioside compositions. Raft analysis revealed that GPI-anchored molecules such as DAF and NCAM tended to disappear from the raft fraction with intensities of DKO > GM2/GD2 synthase KO > GD3 synthase KO > WT. Even lipid raft markers tended to disperse from the raft fractions with similar intensities. Phospholipids and cholesterol also tended to decrease in GEM/rafts, although total amounts were almost equivalent. All these results indicate that GEM/rafts architecture is destroyed by lack of gangliosides with gradual intensity depending on the degree of defects of their compositions.

10. Knockout Mice of Neutral Glycolipid Synthases

As for glycosyltransferase genes responsible for the synthesis of neutral glycolipids, Gb3/CD77 synthase [43] and amino-CTH synthase (β3GnT5) [44] have been genetically disrupted. The α1,4-galactosyltransferase gene responsible for the synthesis of globotrialsylceramide (Gb3) has been isolated in three laboratories [45-47]. It is a key enzyme in the initiation of the synthesis of globo-series glycolipids. This antigen has been considered to be a receptor for Verotoxins from O157 pathogenic *E. coli.* The Gb3/CD77 synthase KO mice showed no definite abnormal phenotypes [43]. However, the mutant mice showed clear differences in the response to injected verotoxins, i.e. they were completely resistant to verotoxins from *E. coli* O157. These results indicated that Gb3/CD77 is a sole receptor for the toxins, and Gb3/CD77 on endothelial cells were the target for verotoxins.

In the analysis of the sensitivity of Gb3/CD77 synthase KO mice to lipopolysaccharides (LPS), they showed increased sensitivity to injected LPS, suggesting globoseries-GSLs might exert to protect cells from LPS toxicity [48]. Consequently, it was demonstrated that globotetraosylceramide (Gb4) specifically binds LPS receptor TLR-4/MD-2, resulting in the competitive inhibition with LPS in the binding to its receptor. As it is well known that LPS binds its receptor TLR-4/MD-2 in lipid rafts, TLR-4/MD-2 moved to lipid rafts where Gb4 persistently resides after LPS stimulation. Moreover, it was shown that Gb4 suppressed LPS toxicity by injection after LPS administration, suggesting that Gb4 can be used therapeutically for endotoxic shock.

11. Implication of GSLs in the Regulation of Architecture and Function of Lipid Rafts

In these studies using KO mice of various glycosyltransferase genes, implications of GSLs in the suppression of inflammation and neurodegeneration were clearly demonstrated. These features in nervous tissues observed in KO mice seem similar to neuroinflammation in human neurodegenerative diseases such as Alzheimer disease and Parkinson's disease [49, 50]. As shown in triple KO mice lacking two ganglioside synthase genes and C3 component of the complement system, immune suppression sometimes caused alleviation of pathological and clinical features in these neurodegenerative diseases.

What should we draw from results obtained from KO mice analysis? The majority of studies on lipid rafts have been performed using cultured cell lines, and roles of lipid rafts have never been directly analyzed as previously summarized [51]. Many results summarized in this chapter were substantially proved by analyzing molecular reactions in and/near lipid rafts in experimental systems *in vivo* and *in vitro*. Therefore, it might be concluded that the majority of abnormal phenotypes can be explained as results of disrupted architecture and function of lipid rafts due to altered GSLs.

As for modes of action of GSLs in lipid rafts, we would propose two kind of modes for their involvement in the maintenance of structure and function of nervous tissues. First, universal roles to maintain basal environment. Second, specific functions for individual species, which are irreplaceable with other structures. The availability of glycosyltransferase genes and the generation of mutant mice lacking glycosyltransferase genes or their modifying enzymes have been tremendously powerful factors in further promoting research to understand the significance of polymorphic glycolipids in various biological events, such as development, cell growth, differentiation and death. They have also contributed in the analysis of glycolipid function in infection, inflammation, cancer evolution, metastasis, neurodegeneration and regeneration. In fact, many of the new findings obtained from the mutant mice are much more complicated than expected, suggesting that GSLs play roles in a variety of manners. Some of their functions appear to be compensated by similar and dissimilar structures, but others appear to be indispensable. To our surprise, GM3 synthase gene knockout mice [19] showed no apparent morphological and behavioral abnormalities, suggesting that asialo-series gangliosides might be able to compensate for the loss of GM3-derived gangliosides. Combination of various KO mice should be effective for the clarification of functional redundancy among GSLs.

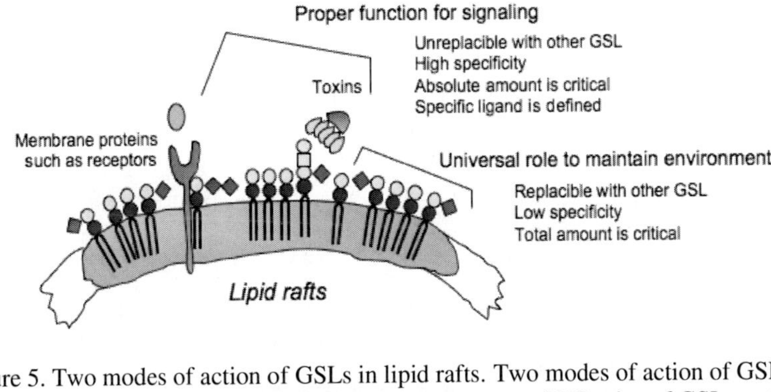

Figure 5. Two modes of action of GSLs in lipid rafts. Two modes of action of GSLs in lipid rafts might be considered to explain the phenotypes of KO mice of GSL synthase genes. One is low-specific and the function is irreplaceable with other GSLs. Therefore, total amounts of GSLs seem to be critical. The other is specific and irreplacible with other GSLs. In this case, specific ligand molecules for the particular structure have been often defined.

On the other hand, specific roles for particular GSLs that can not be compensated have been also identified, e.g. Gb3 as a receptor for Verotoxins [43], or Gb4 as a ligand for TLR-4.MD-2 [48]. Generally, we can expect that these proper functions might be exerted based on the interaction with specific ligand molecules for individual GSLs. Therefore, identification of physiological ligands to individual carbohydrate structures on GSLs is urgently needed to investigate the implication of each specific function. Two modes of action of GSLs in lipid rafts are shown in Fig. 5.

What we should challenge at this time are as follows. It seems essential to investigate formation of molecular clusters between GSLs and/or GSLs and membrane proteins in/near lipid rafts. Enzyme-mediated activation of radical sources (EMARS) reaction seems to be effective to identify interacting molecules in lipid rafts [52, 53]. This approach may be able to allow us to clarify heterogeneity in the architecture and function of lipid rafts. Single molecule imaging with high spatio-temporal resolution should be also a very powerful approach to investigate molecular dynamics of lipid rafts [54, 55].

References

[1] Wiegandt, H., (1985). *Gangliosides* In *Glycolipids*. Ed. Wiegandt, H. Elsevier, Amsterdam, 199-260

[2] Nagata, Y., Yamashiro, S., Yodoi, J., Lloyd, K.O., Shiku, H., Furukawa, K., (1992). Expression cloning of b1, 4 *N*-acetyl-galactosaminyltransferase cDNAs that determine the expression of GM2 and GD2 gangliosides. *J. Biol. Chem.* 267, 12082-12089

[3] Lloyd, K.O., Furukawa, K., (1998). Biosynthesis and functions of gangliosides: recent advances. *Glycoconj J.* 15, 627-636

[4] Furukawa, K., Takamiya, K., Okada, M., Inoue, M., Fukumoto, S., Furukawa, K., (2001). Novel functions of complex carbohydrates elucidated by the mutant mice of glycosyltransferase genes. *Biochim. Biophys. Acta.* 1525, 1-12

[5] Furukawa, K., Tokuda, N., Okuda, T., Tajima, O., Furukawa, K., (2004). Glycosphingolipids in engineered mice: insights into function. *Semin. Cell Dev. Biol.* 15, 389-96

[6] Schengrund, C.L., (1990). The role(s) of gangliosides in neural differentiation and repair: a perspective. *Brain Res Bull* 24, 131-141

[7] Ferrari, G., Fabris, M., Gorio, A., (1983). Gangliosides enhance neurite outgrowth in PC12 cells. *Brain Res.* 284, 215-221

[8] Fukumoto, S., Mutoh, T., Hasegawa, T., Miyazaki, H., Okada, M., Goto, G., et al., (2000). GD3 synthase gene expression in PC12 cells results in the continuous activation of TrkA and ERK1/2 and enhanced proliferation. *J. Biol. Chem.* 275, 5832-5838

[9] Nishio, M., Fukumoto, S., Furukawa, K., Ichimura, A., Miyazaki, H., Kusunoki, S., et al., (2004). Over-expressed GM1 suppresses NGF signals by modulating the intra-cellular localization of NGF receptors and membrane fluidity in PC12 cells. *J. Biol. Chem.* 279, 33368-33378

[10] Ichikawa, S., Sakiyama, H., Suzuki, G., Hidari, K.I., Hirabayashi, Y., (1996). Expression cloning of a cDNA for human ceramide glucosyltransferase that catalyzes the first glycosylation step of glycosphingolipid synthesis. *Proc. Natl. Acad. Sci. U.S.A.* 93, 12654-12658

[11] Yamashita, T., Wada, R., Sasaki, T., Deng, C., Bierfreund, U., Sandhoff, K., et al., (1999). A vital role for glycosphingolipid synthesis during development and differentiation. *Proc. Natl. Acad. Sci. U.S.A.* 96, 9142-9147

[12] Jennemann, R., Sandhoff, R., Wang, S., Kiss, E., Gretz, N., Zuliani, C., et al., (2005). Cell-specific deletion of glucosylceramide synthase in brain leads to severe neural defects after birth. *Proc. Natl. Acad. Sci. U.S.A.* 102, 12459-12464

[13] Bosio, A., Binczek, E., Le Beau, M.M., Fernald, A.A., Stoffel, W., (1996). The human gene CGT encoding the UDP-galactose ceramide galactosyl transferase (cerebroside synthase): cloning, characterization, and assignment to human chromosome 4, band q26. *Genomics* 34, 69-75

[14] Coetzee, T., Li, X., Fujita, N., Marcus, J., Suzuki, K., Francke, U., et al., (1996). Molecular cloning, chromosomal mapping, and characterization of the mouse UDP-galactose:ceramide galactosyltransferase gene. *Genomics* 35, 215-222

[15] Coetzee, T., Fujita, N., Dupree, J., Shi, R., Blight, A., Suzuki, K., et al., (1996). Myelination in the absence of galactocerebroside and sulfatide: normal structure with abnormal function and regional instability. *Cell* 86, 209-219

[16] Fujimoto, H., Tadano-Aritomi, K., Tokumasu, A., Ito, K., Hikita, T., Suzuki, K., et al., (2000). Requirement of seminolipid in spermatogenesis revealed by UDP-galactose: Ceramide galactosyltransferase-deficient mice. *J. Biol. Chem.* 275, 22623–22626

[17] Honke, K., Tsuda, M., Hirahara, Y., Ishii, A., Makita, A., Wada, Y., (1997). Molecular cloning and expression of cDNA encoding human 3'-

phosphoadenylylsulfate:galactosylceramide 3'-sulfotransferase. *J. Biol. Chem.* 272, 4864-4868

[18] Honke, K., Hirahara, Y., Dupree, J., Suzuki, K., Popko, B., Fukushima, K., et al., (2002). Paranodal junction formation and spermatogenesis require sulfoglycolipids. *Proc. Natl. Acad. Sci. U.S.A.* 99, 4227-4232

[19] Yamashita, T., Hashiramoto, A., Haluzik, M., Mizukami, H., Beck, S., Norton, A., et al., (2003). Enhanced insulin sensitivity in mice lacking ganglioside GM3. *Proc. Natl. Acad. Sci. U.S.A.* 100, 3445-3449

[20] Simpson, M.A., Cross, H., Proukakis, C., Priestman, D.A., Neville, D.C., Reinkensmeier, G., et al., (2004). Infantile-onset symptomatic epilepsy syndrome caused by a homozygous loss-of-function mutation of GM3 synthase. *Nat. Genet.* 36, 1225-1229

[21] Kabayama, K,. Sato, T., Saito, K., Loberto, N., Prinetti, A., Sonnino, S., et al., (2007). Dissociation of the insulin *Proc Natl Acad Sci U S A*. 104, 13678-13683

[22] Tokuda, N., Numata, S., Nomura,T., Takizawa, M., Kondo, Y., Yamashita, Y., et al., (2013). b4GalT6 is involved in the synthesis of lactosylceramide with less intensity thanb4GalT5. *Glycobiology* 23, 1175-1183

[23] Kumagai, T., Tanaka, M., Yokoyama, M., Sato, T., Shinkai,.T, Furukawa, K., (2009). Early lethality of b-1,4-galactosyltransferase V-mutant mice by growth retardation. *Biochem Biophys Res Commun.* 379, 456-459

[24] Nishie, T., Hikimochi, Y., Zama, K., Fukusumi, Y., Ito, M., Yokoyama, H., et al., (2010). b4-Galactosyltransferase-5 is a lactosylceramide synthase essential for mouse extra-embryonic development. *Glycobiology.* 20, 1311-1322

[25] Nomura, T., Takizawa, M., Aoki, J., Arai, H., Inoue, K., Wakisaka, E., et al., (1998). Purification, cDNA cloning, and expression of UDP-Gal: glucosylceramide beta-1,4-galactosyltransferase from rat brain. *J. Biol. Chem.* 273, 13570-13577

[26] Takamiya, K., Yamamoto, A., Furukawa, K., Yamashiro, S., Shin, M., Okada, M., et al., (1996). Mice with disrupted GM2/GD2 synthase gene lack complex gangliosides but exhibit only subtle defects in their nervous system. *Proc. Natl. Acad. Sci. U.S.A.* 93, 10662-10667

[27] Takamiya, K., Yamamoto, A., Furukawa, K., Zhao, J., Fukumoto, S., Yamashiro, S., et al., (1998). Complex gangliosides are essential in spermatogenesis of mice: possible roles in the transport of testosterone. *Proc. Natl. Acad. Sci. U.S.A.* 95, 12147-12152

[28] Sugiura, Y., Furukawa, K., Tajima, O., Mii, T., Honda, T., Furukawa, K., (2005). Sensory nerve-dominant nerve degeneration and remodeling in the mutant mice lacking complex gangliosides. *Neuroscience* 135, 1167-1178

[29] Sheikh, K.A., Sun, J., Liu, Y., Kawai, H., Crawford, T.O., Proia, R.L., et al., (1999). Mice lacking complex gangliosides develop Wallerian degeneration and myelination defects. *Proc. Natl. Acad. Sci. U.S.A.* 96, 7532-7537

[30] Chiavegatto, S., Sun, J., Nelson, R.J., Schnaar, R.L., (2000). A functional role for complex gangliosides: motor deficits in GM2/GD2 synthase knockout mice. *Exp. Neurol.* 166, 227-234

[31] Kittaka, D., Itoh, M., Ohmi, Y., Kondo, Y., Fukumoto, S., Urano, T., et al., (2008). Impaired hypoglossal nerve regeneration in complex ganglioside-lacking mutant mice: down-regulation of neurotrophic factors and receptors as possible mechanisms. ***Glycobiology*** 18, 509-516

[32] Wu, G., Xie, X., Lu, Z.H., Ledeen, R.W., (2001). Cerebellar neurons lacking complex gangliosides degenerate in the presence of depolarizing levels of potassium. *Proc, Natl, Acad, Sci, U.S.A.* 98, 307-312

[33] Kawai, H., Sango, K., Mullin, K.A., Proia, R.L., (1998). Embryonic stem cells with a disrupted GD3 synthase gene undergo neuronal differentiation in the absence of b-series gangliosides. *J. Biol. Chem.* 273, 19634-19638

[34] Okada, M., Itoh, M., Haraguchi, M., Okajima, T., Inoue, M., Ohishi, H., et al., (2002). b-series ganglioside deficiency exhibits no definite changes in the neurogenesis and the sensitivity to Fas-mediated apoptosis, but impairs regeneration of the lesioned hypoglossal nerve. *J. Biol. Chem.* 277, 1633-1636

[35] De Maria, R., Lenti, L., Malisan, F., d'Agostino, F., Tomassini, B., Zeuner, A., et al., (1997). Requirement for GD3 ganglioside in CD95 *Science* 277, 1652-1655

[36] Ito, M., Fukumoto, S., Iwamoto, T., Mizuno, A., Rokutanda, A., Ishida, H.K., et al., (2001). Specificity of carbohydrate structures of gangliosides in the activity to regenerate the rat axotomized hypoglossal nerve. *Glycobiology* 11, 125-130

[37] Inoue, M., Fujii, Y., Furukawa, K., Okada, M., Okumura, K., Hayakawa, T., et al., (2002). Refractory skin injury in the complex knock-out mice expressing only GM3 ganglioside. *J. Biol. Chem.* 277, 29881-29888

[38] Kawai, H., Allende, M.L., Wada, R., Kono, M., Sango, K., Deng, C., et al., (2001). Mice expressing only monosialoganglioside GM3 exhibit lethal audiogenic seizures. *J. Biol. Chem.* 276, 6885-6888

[39] Ohmi, Y., Tajima, O., Ohkawa, Y., Mori, A., Sugiura, Y., Furukawa, K., et al., (2009). Gangliosides play pivotal roles in the regulation of complement systems and in the maintenance of integrity in nerve tissues. *Proc. Natl. Acad. Sci. U.S.A.* 106, 22405-22410

[40] [40] Ohmi, Y., Tajima, O., Ohkawa, Y., Sugiura, Y., Furukawa, K., Furukawa, K., (2011). Gangliosides are essential in the protection of inflammation and neurodegeneration via maintenance of lipid rafts: elucidation by a series of ganglioside-deficient mutant mice. *J. Neurochem.* 116, 926-935

[41] Yamashita, T., Wu, Y.P., Sandhoff, R., Werth, N., Mizukami, H., Ellis, J.M., et al., (2005). Interruption of ganglioside synthesis produces central nervous system degeneration and altered axon-glial interactions. *Proc. Natl. Acad. Sci. U.S.A.* 102, 2725-2730

[42] Ohmi, Y., Ohkawa, Y., Yamauchi, Y., Tajima, O., Furukawa, K., Furukawa, K., (2012). Essential roles of gangliosides in the formation and maintenance of membrane microdomains in brain tissues. *Neurochem. Res.* 37, 1185-1191

[43] Okuda, T., Tokuda, N., Numata, S., Ito, M., Ohta, M., Kawamura, K., et al., (2006). Targeted disruption of Gb3/CD77 synthase gene resulted in the complete deletion of globo-series glycosphingolipids and loss of sensitivity to verotoxins. *J. Biol. Chem.* 281, 10230-10235

[44] Togayachi, A., Kozono, Y., Ikehara, Y., Ito, H., Suzuki, N., Tsunoda, Y., et al., (2010). Lack of lacto/neolacto-glycolipids enhances the formation of glycolipid- enriched microdomains, facilitating B cell activation. *Proc Natl Acad Sci U S A.* 107, 11900-11905

[45] Keusch, J.J., Manzella, S.M., Nyame, K.A., Cummings, R.D., Baenziger, J.U., (2000). Cloning of Gb3 synthase, the key enzyme in globo-series glycosphingolipid synthesis, predicts a family of alpha 1, 4-glycosyltransferases conserved in plants, insects, and mammals. *J. Biol. Chem.* 275, 25315-25321

[46] Kojima, Y., Fukumoto, S., Furukawa, K., Okajima, T., Wiels, J., Yokoyama, K., et al., (2000). Molecular cloning of globotriaosylceramide/CD77 synthase, a glycosyltransferase that initiates the synthesis of globo series glycosphingolipids. *J. Biol. Chem.* 275, 15152-15156

[47] Steffensen, R., Carlier, K., Wiels, J., Levery, S.B., Stroud, M., Cedergren, B., et al., (2000). Cloning and expression of the histo-blood group Pk UDP-galactose: Galbeta-4Glcbeta1-cer alpha1, 4-galactosyltransferase. Molecular genetic basis of the p phenotype. *J. Biol. Chem.* 275, 16723-16729

[48] Kondo, Y., Tokuda, N., Nishitani, C., Ohto, U., Akashi-Takamura, S., Ito, Y., et al., (2013). TLR4-MD-2 complex is negatively regulated by an endogenous ligand, globotetraosylceramide in vascular endothelial cells. *Proc. Natl. Acad. Sci. U.S.A.* 110, 4714-4719

[49] Rogers, J., Cooper, N.R., Webster, S., Schultz, J., McGeer, P.L., Styren, S.D., et al., (1992). Complement activation by beta-amyloid in Alzheimer disease. *Proc Natl Acad Sci U.S.A*. 89: 10016-10020

[50] Shen, Y., Lue, L., Yang, L., Roher, A., Kuo, Y., Strohmeyer, R., et al., (2001). Complement activation by neurofibrillary tangles in Alzheimer's disease. *Neurosci Lett* 305, 165-168

[51] Furukawa, K., Tajima, O., Okuda, T., Tokuda, N., Furukawa, K., (2007). *Knockout mice and glycolipids*, In *Comprehensive glycoscience*, From *Chemistry' to Systems Biology*, Eds. Kamerling, J.P., Boons, G.J., Lee,Y.C., Suzuki,A., Taniguchi,N., Voragen, A.G.J. Elsevier, Oxford, UK, 149-157

[52] Kotani, N., Gu, J., Isaji, T., Udaka, K., Taniguchi, N,, Honke, K., (2008). Biochemical visualization of cell surface molecular clustering in living cells. *Proc. Natl. Acad. Sci. U.S.A.* 105, 7405–7409

[53] Hashimoto, N., Hamamura, K., Kotani, N., Furukawa, K., Kaneko, K., Honke, K., et al., (2012). Proteomic analysis of ganglioside-associated membrane molecules: Substantial basis for molecular clustering. *Proteomics* 12, 3154-3163

[54] Suzuki, K.G., Kasai, R.S., Hirosawa, K.M., Nemoto, Y.L., Ishibashi, M., Miwa, Y., et al., (2012). Transient GPI-anchored protein homodimers are units for raft organization and function. *Nat. Chem. Biol.* 8, 774-778

[55] Suzuki, K.G., Kasai, R.S., Fujiwara, T.K., Kusumi, A., (2013). Single-molecule imaging of receptor-receptor interactions. *Methods Cell Biol.* 117, 373-390

In: Lipid Rafts
Editor: Dan Sillence
ISBN: 978-1-62948-961-2

Chapter II

Biophysics of Lipid Rafts and Their Interplay with Ceramide: Studies in Model Systems and Biological Insights

Rodrigo F. M. de Almeida[1], Joaquim T. Marquês[1] and Liana C. Silva*[2]

[1] Centro de Química e Bioquímica, Faculdade de Ciências, Universidade de Lisboa, Lisboa, Portugal

[2] iMed.UL - Research Institute for Medicines and Pharmaceutical Sciences, Faculdade de Farmácia, Universidade de Lisboa, Lisboa, Portugal

Abstract

It is now recognized that lipids participate actively in cellular processes as second messengers, by interacting specifically with certain proteins and/or by providing specialized membrane regions presenting

* Corresponding author: Liana C. Silva, iMed.UL - Research Institute for Medicines and Pharmaceutical Sciences, Faculdade de Farmácia, Universidade de Lisboa, Av. Prof. Gama Pinto, 1649-003 Lisboa, Portugal. Tel: + 351 217 946 400 (ext 14204); Fax: + 351 217 937 703; E-mail: lianacsilva@ff.ul.pt.

distinct biophysical properties. Several types of such membrane domains have been identified, as lipid raft domains and ceramide-platforms.

Lipid rafts in mammalian cells are sphingolipid (mainly sphingomyelin [SM])/Cholesterol (Chol)-enriched domains in coexistence with regions where the lipids are more disordered due to the abundance of unsaturated species, mainly phosphatidylcholine (PC) but also other phospholipids.

Therefore, cell membranes are biophysically described as a fluid or liquid disordered phase (l_d) in coexistence with liquid ordered phase (l_o) domains which would correspond to the lipid rafts. This is recognizably a simplification of the biological system, with a large variety of lipids and proteins and thus of membrane microdomains. Nonetheless, the l_d/l_o region of ternary PC/SM/Cholesterol phase diagrams giving the composition of lipid rafts has been extremely useful. Phase diagrams allow determining the type, number, fraction, composition and size of lipid phases and domains in multicomponent model systems, giving important insights into the behavior of biological membranes and helping the design and interpretation of experiments involving them. In addition, the l_d/l_o partition of probes, proteins and drugs has been quantified using those diagrams.

Ceramide-platforms are thought to exist in a state closer to the gel or solid ordered phase (s_o). To date there is no direct evidence for its existence in mammalian cells under physiological conditions. Nonetheless, in stress situations, where ceramide levels can increase several-fold by the action of sphingomyelinase (SMase), a profound reorganization of the membrane is expected, as a consequence of the strong alterations in membrane biophysical properties that take place. Protein redistribution and changes in activity may occur. Studies in model systems have shown that SM and ceramide act synergistically in the formation of highly rigid and ordered gel domains. On another hand, it was demonstrated that Chol in high amounts is able to solubilize those domains. Thus, a subtle interplay between Chol/SM/Ceramide amounts may dictate at large lipid microdomain organization in mammalian membranes; it is predicted that inner membrane lipids are organized very differently from the plasma membrane. It has been shown that sphingomyelinase action has strong effects on membrane organization which can be rationalized with the phase diagrams and quantitative description of PC/Chol/SM/Ceramide systems. Finally, sphingolipid acyl chain length and degree of unsaturation, which are tissue-specific and change with the state of the cell, also affect membrane biophysical properties.

Keywords: Lipid rafts, ceramide-platforms, liquid-ordered domains, gel domains, phase diagrams, membrane biophysics

Abbreviations

AFM – Atomic Force Microscopy
DOPC - 1,2-dioleoyl-*sn*-glycero-3-phosphocholine
DPH – Diphenylhexatriene
DPPC – 1,2-dipalmitoyl-*sn*-glycero-3-phosphocholine
EPR – Electron Paramagnetic Resonance
FCS – Fluorescence Correlation Spectroscopy
FRAP – Fluorescence Recovery After Photobleaching
FRET – Förster's Resonance Energy Transfer
GPMV – Giant Plasma Membrane Vesicles
GUV – Giant Unilamellar Vesicles
l_d - Liquid disorded
l_o - Liquid ordered
NMR – Nuclear Magnetic Resonanse
PC – Phosphatidylcholine
PCer – N-palmitoyl-D-*erythro*-sphingosine
PLAP – Placental Alkaline Phosphatase
POPC – 1-palmitoyl-2-oleoyl-*sn*-glycero-3-phosphocholine
PSM – N-palmitoyl-D-*erythro*-sphingosylphosphorylcholine
SM – Sphingomyelin
SMase – Sphingomyelinase
SL – Sphingolipid
s_o - Solid ordered
STED – Stimulated Emission Depletion
T_m – Main gel-to-fluid phase transition temperature
t-PnA – trans-parinaric acid

1. Introduction

Over the last decades it became clear that the complex lipid composition of biological membranes is related to their propensity to form distinct types of domains that together with protein form membrane microdomains with specific functions. Both lipid-lipid interactions and the segregation of lipids into domains with specific properties in terms of rigidity, size or thickness confer to biomembranes unique characteristics that allow them to play their role both as selective barriers and platforms where a myriad of relevant physiological processes take place. The structural diversity of lipids in terms of their polar head group and/or the length and unsaturation degree of hydrocarbon chains may drive the lateral segregation of lipids into domains which are believed to be one of the mechanisms of cellular lipid and protein trafficking and sorting. Not only are lipid species important per se, as they can interact specifically with certain proteins being substrates, activators or inhibitors, but most importantly the biophysical properties that they impart to determined membrane regions are crucial for biological function and associated regulatory mechanisms. Of special relevance are the lipid domains known as lipid rafts which have been the most extensively explored type of lipid domains due to all biological significance associated to this kind of naturally occurring structures. In addition to lipid rafts, Ceramide-platforms have also emerged as an important type of lipid domains, which are thought to be formed upon hydrolysis of SM within the lipid rafts, and responsible for the clustering of receptors and consequent activation of signalling events. Understanding the interplay between the lipids involved in the formation of lipid rafts and ceramide-platforms is thus fundamental to comprehend their biophysical relevance and their biological implications.

2. Lipid Raft Domains

2.1. The Lipid Raft Hypothesis and the Liquid Ordered Phase of Membranes

In the same decade in which the fluid mosaic model for the structure of biomembranes was proposed by Singer and Nicholson [1] the concept of membrane microdomains was starting to emerge [2,3]. In the following decades it was realized that the plasma membrane of mammalian cells is compartmentalized in regions with different composition and properties [4,5], and this concept is now extended to other eukaryotic organisms, including plants and fungi [6], as well as to prokaryotes [7]. The nature of such lipid domains has been the subject of intense research, and after long controversy, it was established that a complete description of cell membranes should contemplate the possibility of the coexistence of multiple types of domains that cannot be all described under the same basic concept.

The study of cellular traffic and sorting in polarized epithelial cells, where two clearly distinct membrane regions exist and is steadily maintained by gap junctions, the apical and the basolateral domains, established in the 80s that not only proteins but also lipids play active roles in the process of domain formation [8]. The lipid composition of basolateral and apical membranes was analysed, and the major result was that the latter was enriched in cholesterol and sphingolipids as compared to the former which had a higher content of glycerophospholipids. In terms of acyl chain composition, there was also a significant difference, because sphingolipids are acylated with longer and more saturated acyl chains. Therefore, the basolateral membrane is enriched in acyl chains that on average are shorter and more unsaturated than the apical membrane. At this time, and even presently, detergent-based methods have been very important in the purification and solubilization of membrane proteins [9,10]. An important observation for the membrane domain field was that not all the membrane was solubilized in such extraction procedure, which was usually accomplished using non-ionic detergents, such as TX-100 at low temperature (4°C). An analysis of the composition of the non-soluble fraction revealed enrichment in certain proteins, and most strikingly, in certain lipids, particularly cholesterol and saturated lipids, such as glycosphingolipids [11]. It was also known from the biophysical point of view, that saturated lipids, particularly choline-based lipids phosphatidylcholine (PC) and sphingomyelin (SM), in the presence of cholesterol are able to form a l_o phase, which will be describe in detail below. Moreover, reconstitution of a GPI-anchored protein in model systems in this l_o phase rendered the protein detergent insoluble, whereas reconstitution into a "normal" fluid membrane made of e.g. egg PC, did not [12]. These key findings, together with a wealth of other studies laid the foundations of the lipid raft hypothesis, i.e. along the membrane traffic pathway sphingolipids and cholesterol interact strongly between them, segregating from the remainder of the fluid membrane into regions that are on average more ordered and compact. Certain proteins would be preferentially incorporated into these regions, whereas others would be excluded. This model advanced by Simons and Ikonen [13] provided a framework to address the subject of membrane microdomains and explained in a simple way a large amount of experimental results available at the time. Currently, it is recognized that detergent resistant membranes have little, if any, relation with possible membrane micro or nanodomains pre-existent in the membrane of the cells from which they were obtained. The limitations of procedures which involve detergent extraction to study lipid rafts have been extensively discussed and reviewed (e.g., [6,14,15]. However, the idea that certain lipids have stronger interactions and consequently tend to segregate thereby forming or contributing to the formation of specialized membrane regions with specific functional roles is still accepted by the biochemical and biophysical communities [5,16].

Membrane microdomains in plant and fungi (cell-walled eukaryotes), even those that co-localize with ergosterol and are thought to be sphingolipid-enriched, have many features that are distinct from those usually assigned to lipid rafts in mammalian cells [6]. Perhaps the most striking is the existence of spatio-temporally stable membrane compartments, which drastically differ from mammalian lipid rafts, which are currently defined as nanodomains, highly dynamic, and of small dimensions, but which coalesce to form larger signaling platforms upon certain stimuli, such as receptor cross-linking, external signals or membrane traffic [17]. In this chapter, we will restrict ourselves to cholesterol-enriched lipid rafts, i.e., those occurring in mammalian cells. Other types of membrane microdomains, such as sphingolipid-enriched domains, or lipid lateral heterogeneity that cannot be described by the dichotomist view of l_o/l_d coexistence underlying biophysically the "conventional" lipid rafts are out of the scope of this review and are reviewed elsewhere [18,19]. The exception are ceramide-enriched platforms, which will be the subject of the second part of this review, due to their intimate relation to lipid rafts on one hand, and to the importance of the interplay between these two types of domains in membrane organization and cellular function, on the other.

2.2. Lipid Raft Composition – Binary and Ternary Phase Diagrams

One of the key biophysical concepts in the establishment of the lipid raft as a membrane organizing principle was the discovery of the l_o phase. A biophysical property of cholesterol which has been considered as one of its most important characteristic is the ability to abolish the gel-fluid phase transition of phospholipids, by provoking the occurrence of an intermediate state of lipid bilayers called the liquid ordered phase [20]. Cholesterol is a major component of the mammalian plasma membrane and is required for normal cell growth and function [21]. Therefore the interactions of phospholipids with cholesterol have been intensively studied, and it was established that the biophysical properties of a membrane enriched in cholesterol justify by large its abundance and importance for animal life. For example, cholesterol is able to increase the mechanical strength and decrease the cross-membrane permeability to small molecules and ions of fluid bilayers.

Many of the complex effects of cholesterol in phospolipid bilayers, which were also observed in more complex artificial membranes composed of multiple lipids or even those obtained from natural sources, could be rationalized considering that in presence of cholesterol a new lipid phase is formed. This l_o phase, with properties midway between the gel and the fluid, was first described in binary PC/Chol mixtures. According to the nomenclature introduced by Ipsen et al. [22], the gel and fluid phases are designated by solid ordered (s_o) and liquid disordered (l_d), respectively. The phase diagram proposed is monotectic and for intermediate cholesterol concentrations phase coexistence occurs: s_o and l_o, below the monotectic temperature (which is close to main gel/fluid phase transition temperature, T_m) and l_d with l_o, above the monotectic temperature. Different spectroscopic and calorimetric methods have been employed to determine such phase diagrams. Recently, these diagrams have been collected, compared and critically commented [23]. In that review, it is also explained how to interpret such phase diagrams, how to calculate the l_o fraction and composition for a given temperature and phospholipid/sterol proportion.

Taking into account that lipid rafts are sphingolipid/cholesterol enriched domains, they have thus been considered the biological counterpart of l_o domains /phase in membrane model systems. The most important region of the PC/cholesterol phase diagrams in the context of the raft hypothesis is the l_o/l_d phase separation region and in fact, the proposal of phase diagrams with the coexistence of liquid phases provided the biophysical basis for the existence of lateral heterogeneity arising from specific lipid-lipid interactions, and that was still compatible with the widely accepted fluid nature of biological membranes. This would be in contrast with the other type of binary lipid phase diagrams, for a mixture of two phospholipids, or a phospholipid and a sphingolipid, in which the most important phase coexistence region is that of gel/fluid, not uncommonly found are also regions of gel/gel immiscibility, but usually these occur at temperatures that are considered too low to be relevant in the physiological context, and only very rarely, fluid/fluid phase separation has been proposed [24]. Importantly, in the case of phosphatidylethanolamines/PC mixtures, it is possible to find phase diagrams with coexistence of lamellar and hexagonal phases.

Although phase diagrams of binary mixtures showing l_o/l_d coexistence were important for the formulation of the lipid raft hypothesis, biophysicists soon realized that, in order to understand the formation of raft-like domains, such diagrams were not sufficient because, in addition to a sterol, the model membranes should contain both a high T_m lipid (that would correspond to the sphingolipids in the lipid raft) and a low T_m lipid (that would correspond to the largely unsaturated glycerophospholipids forming the more disordered sea where the rafts were floating [25,26]). For such studies, the low T_m lipid used was almost always an unsaturated phosphatidylcholine, whereas the high T_m lipid could be either a saturated PC or a sphingolipid, which then was usually sphingomyelin.

Thus, lipid phase diagrams contributed to the raft hypothesis and this one stimulated the study of more complex lipid mixtures and the establishment of other phase diagrams. The ternary lipid phase diagrams are usually determined at constant temperature, and are represented in a triangle, the Gibbs diagram, where all the proportions between the 3 components can be mapped. More recently, three-dimensional ternary diagrams where temperature is also a variable have been determined (e.g, [27]).

One of the major advantages of ternary lipid mixtures such as those described above is that micrometer-scale domains could be clearly observed by conventional fluorescence microscopy. On another hand, in binary PC/Chol mixtures, those domains have not been observed, and as pointed out by Marsh [23], ternary mixtures offer more versatile conditions for raft formation than do mixtures of a single lipid with cholesterol. The observation of coexisting l_d and l_o phases in giant unilamellar vesicles (GUV) generated using lipids that were either synthetic or extracted from mammalian cells was an important step in the biophysics of lipid rafts [28].

In the comprehensive review by Marsh [25], a collection of ternary phase diagrams is presented. It is explained how to read and interpret such phase diagrams, and the controversies and differences for diagrams retrieved by different techniques, highlighting the common features and those that can be considered solidly established are given. One of the important characteristics to take into consideration is that to build a ternary phase diagram, it is necessary to know the behavior of the three binary combinations of lipids: low T_m/cholesterol; high T_m/ cholesterol, low T_m /high T_m. In this context, knowledge about the gel/fluid phase coexistence region for binary lipid mixtures regains importance. Here we will consider component 1 as the low T_m lipid, component 2 the high T_m lipid and cholesterol as component 3. We will focus on the region around the 1:1:1 equimolar mixture of those components, which has been considered the canonical raft mixture, and representative of the lipid composition of mammalian plasma membrane [29]. This is particularly true when the high T_m lipid is sphingomyelin, and the low T_m lipid is a 1-saturated,2-unsaturated phosphatidylcholine. In many of the phase diagrams available in the literature, this composition lays either in the single phase l_o region or within a broad l_d/l_o phase coexistence region, and therefore may be considered representative of the lipid raft organization of cell membranes [25].

2.3. Interpreting a Ternary Phase Diagram: Composition and Size of Rafts and Effects of an Anti-Cancer Compound

The first phase diagram determined for the mixture 1-palmitoyl-2-oleoyl-*sn*-glycero-3-phosphocholine (POPC) / *N*-palmitoyl-D-*erythro*-sphingosylphos phorylcholine (PSM) / Chol was obtained from a set of steady-state and time-resolved data, including fluorescence anisotropy and dynamic quenching, and available literature data (Nuclear Magnetic Resonanse (NMR), calorimetry, X-ray diffraction and Atomic Force Microscopy (AFM)) on the 3 binary and also on the ternary system and analogue systems with slight differences in lipid structure [30]. The tie-line containing the 1:1:1 composition was also estimated at 23ºC and 37ºC, and from this it is possible to calculate for any ternary mixture along that line the fraction of the l_d and l_o domains. The composition of the phases is given by the two extremes of the tie-line. The uncertainty of the tie-line at 23ºC was further reduced by additional experimental measurements [31]. In Table 1 these are shown together with exemplificative points along the line, and also results from detergent extraction, and more recently, the approximate compositions of lipid extracts obtained from cancer cells, treated and untreated with the anticancer drug 2-hydroxyoleic acid (2OHOA).

The main conclusions obtained from the tie-line are the enrichment in both sphingomyelin and cholesterol when going from the l_d to the l_o phase. Moreover, there is a striking similarity of the composition of these phases with what was obtained from TX-100 detergent extraction at 4ºC, for the detergent soluble and insoluble fraction, respectively [12]. This result was interpreted as a coincidence, since the extraction was performed at a much lower temperature. It was considered that the apparent agreement was an intricate result of the contribution of both perturbation of the membrane by the detergent on one hand, and the effects of lowering the temperature on another [30].

As stated above, the knowledge of the l_d-l_o tie-line allows choosing compositions where the fraction of each phase is known. In the table two examples are given, and it can be concluded, for example that at 23ºC, the 1:1:1 equimolar mixture corresponds to a membrane that is mostly in the l_o

phase. By preparing a series of samples with compositions along a tie-line it is possible to study how several membrane properties change from the l_d to the l_o phase. One important application is the determination of l_d/l_o partition coefficients of membrane probes. This in turn is essential for certain studies, such as Förster's Resonance Energy Transfer (FRET) with fluorescently labeled lipids [31,34]. One such study was performed in order to assess the changes of size undergone by l_o raft-like domains along this tie-line.

Table 1. Ternary mixtures of POPC, PSM and Chol with the given molar fraction (*X*) of each component along the tie-line that contains the 1:1:1 equimolar "canonical" raft mixture and respective l_o phase fraction calculated from the estimated tie-lines taken from the ternary phase diagrams at 23°C and 37°C [30,31]. The detergent soluble and insoluble fractions are the results of TX-100 extraction at 4°C from Schroeder et al. [12], and in this case bovine liver instead of POPC and bovine brain SM instead of PSM were used. The l_o composition at 37°C is comparable to the TX-100 insoluble fraction and the l_d (0% l_o) composition at the same temperature is comparable to the soluble fraction of the initial liposomes used in the detergent extraction study. The human glioma U-118 cell composition was obtained after lipid extraction using a modified n-hexane/2-propanol (3:2, by volume) extraction method [32], followed by separation by high performance thin layer chromatography and phosphate quantification [33]. Cell treatment was carried out with addition of 200 μM of 2-hydroxyoleic acid to the culture medium followed by incubation for 48 hours [33]. See text for the approximation of lipid extract composition to ternary POPC:PSM:Chol mixtures

System	X_{POPC}	X_{PSM}	X_{Chol}	X_{lo}
Tie-line at 23°C	71.6	23.3	5.0	0
Tie-line at 23°C	25.4	34.8	39.8	100
Tie-line at 23°C	59.7	26.3	14.0	26
Tie-line at 23°C	33.3	33.3	33.4	83
Tie-line at 37°C	69 – 57	18 – 33	13 – 10	0
Tie-line at 37°C	14 -19	42 – 33	44 – 48	100
Detergent soluble fraction	60	26	14	------
Detergent insoluble fraction	16	38	46	------
Control U-118 cells	51.2	19.9	28.9	54
Treated U-118 cells	39.8	34.3	25.9	67

In such study it was concluded that along the tie-lie, as the fraction of the l_o domains increase, also their size becomes larger. It was also concluded that the maximum size of the domains becomes larger when more sphingomyelin is present [31].

Another application of the POPC/PSM/Chol phase diagram was the study of the biophysical effects of treating human glioma cells with the anti-cancer drug 2-hydroxyoleic acid [33]. The lipid composition of control cells was determined as well as those of cells treated with the drug at different concentrations and challenging times. To model the lipid components of those cell membranes, only the fractions of SM, PC, cholesterol and PE were taken into account, since there were only marginal changes in phosphatidylserine levels, and phosphatidylinositol was always a minor lipid. For the phospholipids, the 1-palmitoyl,2-oleoyl acyl chains were chosen, and for sphingomyelin, the *N*-palmitoyl species was selected as the most representative ones. This choice was validated by the fact that the fluorescence anisotropy of diphenylhexatriene (DPH) and fluorescence lifetime of trans-parinaric acid (t-PnA) measured in the cell lipid extracts reconstituted into liposomes was very similar to those obtained in the membrane model systems [35]. To interpret the changes in lipid composition occurring upon cell treatment with 2OHOA within the framework of the ternary phase diagram, phosphatidylethanolamine was considered as evenly distributed between l_d and l_o phase due to the fact that its T_m is approximately the average of T_m values for POPC and PSM [24]. Again, this approximation was supported by comparing the values of fluorescence anisotropy and lifetime of several membrane probes in the corresponding ternary mixtures [30,31,34]. In Table 1, the compositions are given for control cells and cells treated with 200 μM of 2-OHOA for 48 hours. From the experimental analysis of the membrane models it was possible to concluded that the global order of the U118 cell membranes increased due to a tighter packing in the l_o domains, probably as a consequence of the marked increase in their SM content, revealed by the lipid analysis, and that was in turn due to a strong activation of Sphingomyelin Synthase by the drug [33]. Furthermore, the mapping of the lipid compositions in the ternary phase diagram suggested that the membranes of the treated cells had not only a higher fraction of l_o domains as compared to the control, but that these domains were sphingomyelin-enriched (most of the new sphingomyelin actually should incorporate the raft-like domains), which would then become larger, as predicted by the FRET study previously described [31]. The formation of these larger and tighter l_o domains could provide the biophysical basis for the activation of signaling cascades by 2OHOA observed in other studies, considered key events in its anticancer mechanism of action [36].

2.4. The Properties of Liquid Ordered Domains– Importance of Different Model Systems and Biophysical Techniques

The liquid ordered phase occurs in membrane model systems prepared from a mixture of a saturated) phospholipid or sphingolipid and cholesterol. Therefore, assuming that lipid rafts can be modeled as l_o phase in artificial membrane systems, these can be used to assess a variety of biophysical properties, applying a set of biophysical methods to the study of different types of model systems, where the type of lipids (low T_m and high T_m) and their fraction, as well as the amount of sterol can be changed in a quantitative manner [37]. In the next paragraphs, properties that are benchmarks of the l_o phase will be briefly mentioned. It will also be described in particular cases how they are reflected in experimental data obtained by selected biophysical methods that can be used to study the influence of additional molecules in the model system, such as the proteins, or ceramides, the latter being the scope of the second part of this chapter. It is by no means intended to be an exhaustive compilation of all the studies and properties, but rather a selection of what the authors consider illustrative of their current view on the subject. In most cases the examples given will refer to l_o domain in coexistence with an l_d phase, in ternary lipid mixtures, taken as a valid biophysical model of lipid rafts.

The l_o phase was designated as such because of two fundamental properties. One of them is the high degree of conformational order of the acyl chains and the other is the fast translational mobility of the lipids (low long range order), the former being more similar to the gel phase and the latter to the fluid [20,23,25,29,37]. The high order of the acyl chains has been demonstrated by a multitude of techniques. For example, by fluorescence spectroscopy, the degree of order of the acyl chains can be evaluated from the fluorescence anisotropy of membrane probes with the chromophore buried in the hydrophobic core of the lipid bilayer, such as DPH, or t-PnA. In the PSM/POPC/Chol mixture, for these probes, the fluorescence anisotropy shows a clear increase when going from the l_d to the l_o phase, along a compositional tie-line [30,31,34]. The order parameter profiles obtained by 2H-NMR have revealed this effect for the ternary equimolar mixture at 40ºC along most of the acyl chain NMR (order parameters) [38].

A consequence of the increased order of the membrane throughout most of its depth is a larger thickness of the raft-like domains, due to the increased stretching of the hydrophobic chains. Such result was clearly detected by X-ray diffraction (e.g., [39]). This was also confirmed by AFM studies where the raft like domains can be clearly observed in topographical images as thicker regions of the bilayer (e.g., [40-42]), and has become a very popular method to study raft-like domains [43,44], and their change upon lipid composition alterations or interactions of other molecules with the lipid bilayer. Recently, a method for the determination of quantitative lipid phase diagrams was developed based on AFM imaging.

This method was applied both to binary (1,2-dipalmitoyl-*sn*-glycero-3-phosphocholine (DPPC)/ergosterol) and ternary (1,2-oleoyl-*sn*-glycero-3-phosphocholine(DOPC)/DPPC/ergosterol, complemented with fluorescence microscopy), with the aim of mimicking yeast lipid rafts, and the influence of ethanol, to understand ethanol tolerance in those microorganisms [43,44].

Both membrane order and thickness were presented as key properties that define which proteins are preferentially located in lipid raft domains, particularly, for GPI-anchored and palmitoylated proteins and integral membrane proteins with longer transmembrane domains, respectively [13,45-47].

Another key biophysical feature of l_o raft-like domains is the lateral diffusion coefficient of molecules in this lipid phase. Whereas in a solid ordered or gel state the diffusion coefficient is on the order of 10^{-10} cm^2/s or smaller, in a fluid phospholipid bilayer (i.e., a l_d phase) it is on the order of 10^{-8} cm^2/s. The importance of the l_o phase and the possibility of its coexistence with an l_d phase was, as stated above, that it gives a basis for the formation of ordered domains that retain membrane fluidity. The lateral diffusion coefficient in PC/Chol membranes is reduced only 2 to 3 times for concentrations of cholesterol up to 40% as determined by pulsed field gradient NMR on macroscopically aligned lipid bilayers [48], and by a factor of 3 to 5 in ternary mixture containing sphingomyelin [49]. Several other techniques have been used to study the lateral diffusion of molecules on ternary mixture mimicking raft membranes, and this result is quite general for lipid molecules in an l_o phase [23,25,50]. Another important set of properties that are characteristic of the l_o phase are related to hydration and polarity [51]. One such property membrane dipole potential, which has been suggested to play a major role in the modulation of protein activity, signaling or other raft associated events [52,53]. Such properties have been assessed most commonly through the spectral shifts in membrane probes that are particularly sensitive to their variations [54]. Using polarity-sensitive probes, such as Laurdan and potential sensitive dyes (such as di-8-ANEPPS or di-4-ANEPPDHQ) it has been shown that the l_o phase is less hydrated and has a much higher dipole potential than the l_d phase (e.g. [53]), and the spectral shifts and/or fluorescence lifetime changes of the probes induced by those differences have been used to distinguish l_o from l_d domains in GUVs or even visualize raft-like/ l_o-like regions in cell membranes (e.g., [28]).

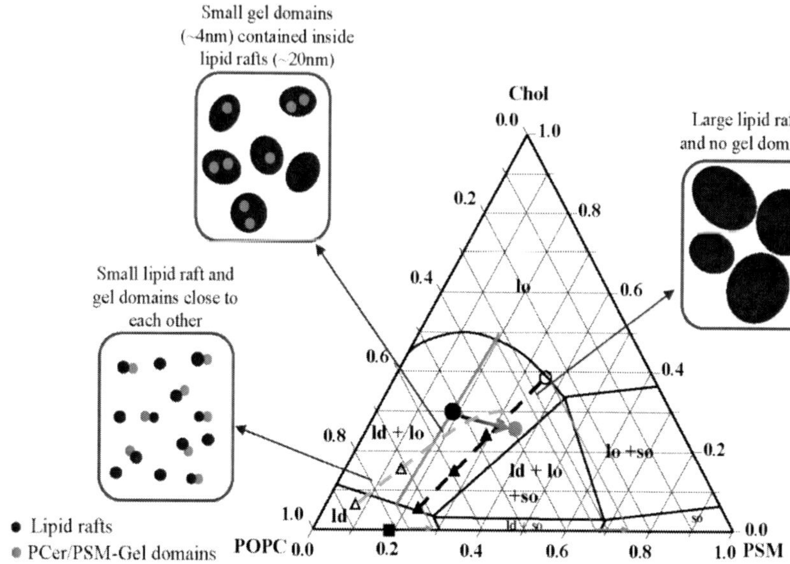

Figure 1. Ternary phase diagram for the POPC, PSM, cholesterol lipid ternary mixture at 23ºC. Several regions of phase coexistence confined by the solid black lines can be depicted in this phase diagram. Of special relevance is the l_o / l_d region that is thought to mimic the presence of lipid rafts. Above a certain cholesterol concentration only one liquid phase is present. Symbols represented along the dashed black tie line correspond to the compositions described in Table 1. This tie line contains the 1:1:1 lipid composition. The representative effects of 2OHOA treatment are shown by the black (control cells) and grey (treated cells) solid circles. A PSM enrichment is observed for both non-raft (small arrow in the PSM axis) and raft (large arrow) fractions of the membrane, especially for the latter. The effects of adding 4 mol% of PCer are also shown. PCer recruits PSM to form gel domains and thus it changes the composition of the mixtures and the position of the tie line of the remaining fluid phase. The dashed grey line and open triangles correspond to the estimated composition of the remaining fluid phase after PSM is recruited for the formation of gel domains. In the exclusively l_d region (solid black square) one molecule of PCer recruits up to 4 molecules of PSM forming highly ordered and very small gel domains. At the low-to-intermediate cholesterol region (solid black triangles), where small lipid rafts are present, PCer/PSM-gel domains are still forming and are surrounding by lipid rafts. In the high cholesterol region (open circle), where there is a predominance of large lipid rafts, PCer/PSM-gel domains are abolished. Adapted from de Almeida et al. 2003 [30]; de Almeida et al. 2005 [31]; Silva et al. 2007 [34]; Barceló-Coblijn et al. 2011 [33].

2.5. Examples of "L_o" Phase Evidence in Biological Systems

Although the simple view of biological membranes as l_o domains in a sea of l_d is currently considered oversimplified, and not accounting for many experimental evidence, it is still a very useful model to rationalize many data (e.g., budding of certain enveloped viruses which have an envelope enriched in sphingolipids and cholesterol as compared to the bulk plasma membrane composition of the host cell[55]). The biological relevance of $l_d - l_o$ model systems is reflected by several results obtained in cells, giant plasma membrane vesicles (GPMV) and reconstituted lipid extracts, some of which referred in the previous section.

GPMV are membrane blebs usually induced by treatment with formaldehyde or a similar compound so-called chemical vesiculants. It is believed that they retain the lipid and protein composition of the original plasma membrane, but without the constraints imposed by interactions with cytoskeleton [56]. Several studies in GMPV, considered as a system closer reflecting the properties of the native plasma membrane of living cells were made, with results pointing towards a behavior consistent with the raft hypothesis, and a $l_d - l_o$ coexistence [57,58].

In all GMPVs analysed by Veatch et al. [59], at a certain range of temperatures $l_d - l_o$ phase separation was observed, and near the temperature at which the membrane undergoes from the two phase to the one phase situation or the reverse, critical behavior was observed, indicating that the composition of the plasma membrane is tuned to be close to a critical composition in a many-dimension phase diagram with fluid-fluid phase coexistence [59]. Since this cannot happens by chance, because critical points are very few in the composition temperature space of multicomponent mixtures thermotropic behavior, the Authors propose that this fine-tuning provides a physical basis for compositional inhomogeneties and other cell membrane properties that are compatible with the definition of lipid rafts.

Evidence for the coexistence of at least two liquid phases with diffusion coefficients of proteins differing by a factor of ~3, as verified in model systems for l_o and l_d, was obtained from FRAP experiments in living epithelial cell lines [60]. Although the long range diffusion of fluorescently-tagged proteins were followed, the recovery patterns were consistent with the existence of two lipid phases containing preferentially different sets of proteins and with one phase being continuous and the other with domains dispersed in a continuous phase, i.e., a percolative structure. Although previous studies have been very important to demonstrate the cholesterol-dependent co-clustering or segregation of proteins in nanodomains < 200 nm (e.g. [61,62]), the FRAP study by Meder et al. presented important support that the co-clustering of proteins into micro/nanodomains in the plasma membrane of living cells was related to lipid phase separation.

One of the problems associated with the detection of lipid rafts in cells is the fact that they are dynamic and small, this meaning with sizes below the resolution of conventional optical microscopy. One way of circumventing these problems is to characterize the complex systems with other spectroscopic properties than solely fluorescence intensity which are able to unravel membrane heterogeneity that is no apparent from that experimental variable alone [50,63]. A few examples have already been described above in this chapter. Selected case studies of membrane domains in vivo using fluorescence lifetimes where the results obtained in living cells are interpreted with the help of results obtained with artificial membranes, reconstituted lipid extracts, isolated plasma membranes and/or GMPVs are explained elsewhere [64].

The development of techniques with high spatial (and temporal) resolution, in particular, the advent of super-resolution nanoscopic techniques have helped/ will help to confirm (or not) key tenets of the raft hypothesis [15,50,65]. For example, one of the most recent and convincing study of lipid rafts in living cells uses the high spatial resolution of stimulated emission depletion (STED) microscopy combined with the high temporal resolution of FCS [66]. This study shows the existence of Chol-dependent membrane domains with sizes around 20 nm where GPI-anchored proteins are transiently trapped in periods of 10-20 ms. One of the most important features in this study is the different behaviour of fluorescently-labeled sphingomyelin versus phosphatidylcholine, where the former is also confined in a similar way as GPI-anchored proteins, but not the latter.

3. Ceramide-Platforms

3.1. Insights from Simplified Model Membranes

It has been postulated that ceramide forms large domains in cell membranes in response to a wide variety of insult, including death receptor ligands, UV-irradiation and chemotherapeutic agents [67-69]. These ceramide-domains, commonly known as ceramide-platforms [70], would derive from the unique biophysical properties of this lipid and would be important modulators of ceramide biological function [71]. Indeed, over the last decades ceramide has emerged as an important sphingolipid (SL) due to its bioactive role in a number of cellular processes and disease [72-74]. The mechanism by which ceramide exerts its biological role is yet not fully known, but accumulating evidence suggests that alterations on membrane biophysical properties derived from increasing levels of ceramide in the plasma membrane might be responsible for the activation/inhibition of ceramide-mediated cellular processes [75-78]. In addition, it has been postulated that the structural features of the ceramide species might control its specificity toward a specific cellular process [79] and this can be due to the distinct changes that each ceramide promotes on cell membrane biophysical properties [80-82].

The biological importance of ceramide, together with the evidence for a role of membrane biophysics in ceramide-mediated biological processes, has stimulated a number of studies that aimed to understand the actual biophysical impact of the different ceramide species [82-98]. Most of our current knowledge on ceramide biophysics results from studies performed in a multitude of model membranes, ranging from monolayers to GUVs, and containing distinct lipid composition. These studies contributed to the understanding of ceramide interplay with other lipids and how this governs membrane biophysical properties, as well as its biological implications. In this section we present a literature review of the most important findings regarding the effects of different ceramides on the biophysical properties of model membranes that contributed to definition of ceramide domains as highly-ordered, tightly-packed ceramide-enriched gel domains.

It is now recognized that ceramide effects on membrane biophysical properties are complex, even when mixed with only one lipid [81,82,97], and strongly dependent on membrane properties and lipid composition [34,80,83,96,98]. This is partially due to the unique molecular structure of Cer, including its very small polar headgroup, strong hydrophobicity and ability to function both as donor and acceptor for H-bonds, creating an extensive H-bonding network that contributes to the high melting temperature [81,99] and in-plane phase separation of ceramide into solid-like condensed domains [81-83,91,97,100,101].

Several studies have addressed the thermotropic and phase properties of distinct ceramide species [76,81,87,99,101,102]. Even though each ceramide has its own thermotropic characteristics, it is collectively reported that ceramides exhibit complex polymorphic phase behaviour [81,99] and display tight packing properties, tighter than other lipids [81,97,103,104]. When mixed with other lipids, ceramide behaviour is even more complex. It has been shown that ceramides have the ability to increase the order of the acyl chains of both saturated [90,105,106] and unsaturated phospholipids [97,107-109] and promote lateral phase separation [90,97,107,110], driving the coexistence of highly-ordered ceramide-enriched gel domains in a ceramide-poor fluid matrix. These features are shared among naturally occurring ceramides with distinct acyl chain structures, albeit significant differences in their effects on membrane properties are observed. We [82] and others [86,93] showed that a decrease in the length and the introduction of an unsaturation in the fatty acyl chain, decreases the ability of ceramide to segregate into gel domains and to increase the packing of the fluid phase [82,86]. In contrast, ceramides with long and very long saturated acyl chains have identical ability to segregate into gel domains and increase the packing of the fluid phase [82]. It has also been reported that the size and morphology of ceramide-gel domains is dependent on ceramide acyl chain structure [82,89,111]. In addition, it has been shown that very long acyl chain ceramides are prone to induced morphological alterations on the membrane, including the formation of tubule-like structures [81,82]. These subtle differences in the behaviour of distinct acyl chain ceramides might be differentially sensed by receptors, proteins and signalling molecules and thus be responsible for the activation of a specific signalling event.

Ceramides are also able to change the properties of membranes in the gel phase and promote gel-gel phase separation when mixed with a variety of phospholipids [90,91,95,107,109,110], as well as with sphingomyelin [85,109,112]. Even though these models are thought to be less representative of biological membranes in the context of the lipid raft hypothesis, they provided important clues to the understanding of ceramide interactions with other lipid components. All of this information has been summarized using phase diagrams similarly to what has been above discussed for the lipid rafts. The phase behaviour of a given ceramide mixed with a particular phospholipid or sphingolipid can thus be obtained directly from the currently available binary phase diagrams [81,82,89,93,97,107,110,113] and used to understand, as well as to predict, its behaviour in more complex systems. This strategy has indeed been utilised to elucidate ceramide interactions with sphingomyelin in mixtures with phospholipids [83,114] and/or cholesterol [34,80,83,96]. It has been shown that ceramide and sphingomyelin have a synergistic action potentiating the formation of highly-ordered ceramide/sphingomyelin-enriched gel domains, both in ternary phospholipid/sphingomyelin/ceramide mixtures [83] and in quaternary phospholipid/sphingomyelin/cholesterol/ceramide mixtures [34,80,96,98]. However, this synergistic effect is highly dependent on the sphingomyelin, ceramide and/or cholesterol content of the mixtures. Accordingly, increased concentrations of sphingomyelin in ternary phospholipid/sphingomyelin/ceramide mixtures promote gel-gel phase separation, similarly to the binary sphingomyelin/ceramide mixtures [85,109,112], rather than a concerted action towards the formation of the ceramide domains.

In mixtures containing sphingomyelin and/or cholesterol the phase behaviour is even more complex, due to the interplay between ceramide and cholesterol [34,80,96,98]. Understanding this interplay has been one of the major challenges among the researchers working in this field. Not long ago, it was passionately debated if i) ceramide and cholesterol compete with each other to associate with sphingomyelin – the ceramide/cholesterol displacement hypothesis [115] or if ii) ceramide-gel domain formation is dependent on the membrane cholesterol content, where ceramide-rich domains are formed at low cholesterol amounts, and abolished when the cholesterol content is increased – the ceramide miscibility in the l_o-phase hypothesis [96]. The displacement hypothesis has been used to justify the results obtained in a number of studies [92,115-120]. However, this hypothesis implies that cholesterol and ceramide are not able to interact, since they would mutually segregate. The evidence for the existence of ceramide/cholesterol domains in monolayers [121,122] and in cell membranes [123,124], including under pathological conditions [125], argues against this hypothesis. The displacement hypothesis has been further challenged by recent studies carried out with phospholipid/ceramide/cholesterol model membranes where the hydrolysis of cholesterol to cholestenone resulted in the appearance of ceramide-gel domains in mixtures where before they were not present [96]. Based on this evidence, it is now becoming accepted that the reported abolishment of ceramide-gel domains [34,80,83,98,111,118,126,127] is a consequence of the increased levels of cholesterol in the mixtures that enhance the miscibility of ceramide in the l_o phase [80,92,96,128]. From a biological perspective, these results are extremely important since it is implicit that ceramide biophysical outcome can be tuned by cholesterol, suggesting that alterations in the cholesterol content of cell membranes might be used by cells to regulate ceramide-mediated processes.

3.2. Interplay between Lipid Rafts and Ceramide-Platforms

Understanding the interplay between ceramide and lipid rafts was a major step towards the recognition of the actual biophysical impact that ceramide can exert on biological membranes. This is because it is widely accepted that, in cell membranes, ceramide is formed by the hydrolysis of sphingomyelin within lipid rafts [75]. Therefore, at the site where ceramide is formed the membrane is enriched in cholesterol and sphingomyelin and should display l_o-like phase properties. As would be anticipated from the discussion presented above, ceramide effects on raft-mimicking membranes are intricate.

Several studies indicate that ceramide can recruit 1 [129] to 3 [34] sphingomyelin molecules from the fluid phase to form a ceramide/sphingomyelin-enriched gel phase [34,126,127,130,131]. The properties of ceramide-gel domains are atypical and even in these complex mixtures ceramide is able to segregate into highly-ordered gel domains [34,80,98,132]. These domains are typically formed within [34,80,96,118] or at the boundaries [133,134] of the l_o phase, suggesting that this fluid ordered phase might function as a protective layer between the gel and the fluid (disordered) phase by decreasing the line tension resultant from gel-fluid phase separation. Nevertheless, ceramide-gel domain formation is abolished in mixtures containing a cholesterol content exceeding that of ceramide, and this was observed both in raft-mimicking mixtures containing pre-mixed ceramide [80,96] and in mixtures where ceramide was generated *in situ* by sphingomyelinase (SMase)-driven sphingomyelin hydrolysis [98,135].

To further illustrate the importance of lipid structure and membrane-induced biophysical changes, one can compare the effects of ceramide and ganglioside GM1 in similar amounts in the same ternary lipid raft mimicking mixture, POPC, PSM, cholesterol. GM1 (from bovine brain, the one used in the study that is being described) has the same Cer backbone as palmitoyl ceramide (PCer), and is acylated mainly with C16:1 and C18 hydrocarbon chains. Moreover, unlike PCer, GM1 contains a bulky and highly hydrated polar head group and tends to form micelles in water. Fluorescence spectroscopy studies have shown the ability of GM1 lipid proportions up to 4 mol% to induce a slight increase in the order of the system, though without gel formation, as well as an enlargement of the l_o domains [31]. Conversely, the presence of PCer up to 4 mol% drives the formation of a gel phase for cholesterol fractions lower than 25 mol%, however it does not promote a change in the size of lipid rafts [34]. Also, in contrast to GM1, PCer leads to a change in the tie line along which the experiments were performed [34] showing that PCer effects are more drastic than the ones induced by GM1. The incorporation of GM1 at 1 mol% in the DOPC, PSM, cholesterol ternary mixture, which exhibits similar phase behaviour to the POPC, PSM, cholesterol mixture, drives the formation of round GM1-enriched microdomains within the l_o domains with sizes ranging from 40 to 150 nm in diameter and 1 nm thicker than the surrounding membrane as revealed by AFM [41]. At a higher proportion of 5 mol%, GM1 forms a net of thin filaments that are generated by the coalescence of the round microdomains detected at a lower GM1 concentration [41]. Also a strong influence in the structure of a DOPC, PSM, cholesterol lipid bilayer was perceived by AFM upon the addition of ceramide. At 5 mol%, ceramide induces an increase of the height difference between l_o and l_d phases, from 1.1 nm to 1.4 nm, even though l_o domains are similar in size and occur as an homogeneous structure [127]. At higher ceramide fractions of 8 and 10 mol% thicker and round-shaped subdomains emerge within the l_o phase and then a three-phase system is formed [126,127].

As above mentioned, the interplay between ceramide and lipid rafts has been studied both i) in premixed ceramide/raft mixtures and ii) *in situ* by SMase-driven SM hydrolysis. Even though the overall conclusions taken from these studies are similar, important differences were noticed when ceramide was formed upon sphingomyelin hydrolysis and which might highlight specific features of ceramide effects in biological membranes. Accordingly, it has been reported that the rapid formation of ceramide in the membrane, by the action of SMase, drives the clustering of domains, which was not observed in premixed raft/ceramide mixtures [98,133,134], suggesting that ceramide-platform formation in cell membranes could be related to the fast conversion of sphingomyelin into ceramide.

The studies where ceramide was generated *in situ* by the action of SMase further revealed that the amount of ceramide formed in the raft-mimicking mixtures has a higher dependency on the initial biophysical properties of the membrane, rather than on the initial content of sphingomyelin [87,98,136,137]. Several studies have highlighted the importance of lipid phase separation on the activity of the enzyme [88,102,126,133,136-140], and it was shown that the activity of SMase was particularly enhanced in mixtures displaying l_o/l_d phase coexistence [98]. From all of these studies it could be concluded that the extent of changes induced by ceramide on membrane biophysical properties are strongly connected to the initial lipid composition and properties of the membrane, which modulate the activity of SMase and thus the amount of ceramide that is generated.

Accordingly, it can be concluded that ceramide will promote greater changes in raft-mimicking membranes, particularly if the cholesterol content of those is low [98]. It can thus be hypothesized that activation of a specific signalling cascade as a consequence of ceramide-platform formation is less likely in a cholesterol-enriched compared to a cholesterol-poor membrane. In this context, it is also important to mention that ceramide-induced alterations are not solely related to the packing properties of the membranes and formation of a gel phase, but also on the structural organization of the membrane [84,113,138,139,141]. It was shown that hydrolysis of sphingomyelin into ceramide leads to vesicular budding and increased permeability of the membranes, which might concertedly act to activate ceramide-mediated cellular processes. It has also been shown that ceramide facilitates the formation of non-lamellar phases [94,95,138,141], and forms interdigitated gel phases [81,82], which seems to enhance the propensity of ceramides to form tubule-like structures.

Such alterations in the packing, lateral and structural organization of the membrane will certainly modulate lipid-lipid and lipid-protein interactions, which will consequently impact on cell function. One of the raised hypotheses to explain how ceramide-induced biophysical changes modulate cellular processes considers that ceramide-domains might be responsible for protein oligomerization and their subsequent partition into or segregation from ceramide-enriched regions. Ceramide-driven reorganization of proteins has been shown using raft-mimicking membranes containing the raft associated GPI anchored protein placental alkaline phosphatase (PLAP) [142]. This protein, initially found in the l_o phase, reallocated into the ceramide-enriched domains upon treatment of the membranes with SMase. Based on this and other [143,144] evidence it was postulated that the tight packing of ceramide-domains might be responsible for the entrapment of receptors and signalling molecules and thus facilitate their interactions. In addition, the creation of packing defects promoted by gel-fluid phase separation is also prone to modulate the activity of interfacial proteins, such as protein phosphatase 2A [105,106], protein kinase C [145] and even SMase [87,98,146].

Conclusion

Even though many studies have been performed to elucidate the characteristics of lipid raft domains and ceramide platforms it is still debated if these domains are formed under biological conditions. Their existence in cellular membranes has for long been questioned, but the current technological development in super-resolution optical microscopy techniques has shed light into this problem (reviewed by [65]). It is now becoming accepted that lipid raft domains are not only domains that result from 'the preferential packing between sphingolipid and cholesterol to constitute platforms onto which proteins can attach' [13] but are 'small (10–200 nm) heterogeneous, highly dynamic, sterol- and sphingolipid-enriched domains that compartmentalise cellular processes. Small rafts can sometimes be stabilised to form larger platforms through protein-protein and protein-lipid interactions' [147].

Despite all the biological implications, studies that aim to characterize the properties of ceramide platforms in the cell context are scarce. In fact, in most of the studies performed in cellular membranes, ceramide-platforms have been identified through the use of antibodies anti-ceramide [148-150]. Information about the properties of ceramide-platforms cannot be obtained in these studies and, in addition, the use of antibodies might even promote the clustering of domains that, otherwise, would not be formed [17,55]. Nevertheless, numerous studies have identified ceramide (-platforms) as required for the activation of several cellular events, including in the clustering and activation of cell death receptors in response to a variety of insult [70,150-155], intracellular trafficking [123,125,156], plasma membrane repair (reviewed in [157]), bacterial [149,158] and viral [155,159] infection and cancer cell death by chemotherapeutic agents and radiation [68,160-162]. The underlying mechanism still requires further elucidation. For example it was determined that the presence of PCer in a POPC, PSM, Chol system is not enough to drive the oligomerization of the transmembrane domain of the Fas death receptor into lipid rafts [144]. So far the most conceivable hypothesis considers that ceramide assists the clustering of receptors while simultaneously recruits the signalling molecules required to the formation of the respective signalling complex [163]. The biophysical evidence obtained from model membrane studies support this hypothesis, inasmuch as ceramides induce strong and diverse changes on membrane properties, in a manner that depends both on the ceramide structure and lipid composition, which can globally account for the varied biological roles of ceramides. A very recent study has provided further support into this hypothesis by showing that generation of ceramides with different acyl chain structures in living HEK cells promotes massive changes on the biophysical properties of cellular membranes that are compatible with the formation of ceramide-enriched gel domains whose characteristics are dependent on the ceramide that is generated [164].

Acknowledgments

Funding from Fundação para a Ciência e Tecnologia, Portugal (PTDC/QUI-BIQ/111411/2009, PTDC/BBB-BQB/0506/2012, SFRH/BD /64442/2009, PEst-OE/QUI/UI0612/2011-2013, Compromisso para a Ciência 2008 and Investigador FCT 2012 (POPH, Fundo Social Europeu) is gratefully acknowledged.

References

[1] Singer, S.J., and Nicolson, G.L. (1972) *Science*, *175*, 720-&
[2] Stier, A., and Sackmann, E. (1973) *Biochim. Biophys. Acta*, *311*, 400-408
[3] Israelachvili, J.N. (1977) *Biochim. Biophys. Acta*, *469*, 221-225
[4] Welti, R., and Glaser, M. (1994) *Chem. Phys. Lipids*, *73*, 121-137
[5] Bagatolli, L.A., Ipsen, J.H., Simonsen, A.C., Mouritsen, O.G. (2010) *Prog. Lipid Res.*, *49*, 378-389
[6] Malinsky, J., Opekarova, M., Grossmann, G., Tanner, W. (2013) *Annu. Rev. Plant Biol.*, *64*, 501-529
[7] Lopez, D., and Kolter, R. (2010) *Gene. Dev.*, *24*, 1893-1902
[8] Simons, K., and van, M.G. (1988) *Biochemistry*, *27*, 6197-6202
[9] Garavito, R.M., and Ferguson-Miller, S. (2001) *J. Biol. Chem.*, *276*, 32403-32406
[10] T. Arnold and D. Linke, The Use of Detergents to Purify Membrane Proteins, Current Protocols in Protein Science, John Wiley & Sons, Inc., 2001.
[11] Brown, D.A. and Rose, J.K. (1992) *Cell, 68*, 533-544.
[12] Schroeder, R., London, E., Brown, D. (1994) *Proc. Natl. Acad. Sci. USA*, *91*, 12130-12134
[13] Simons, K., and Ikonen, E. (1997) *Nature*, *387*, 569-572
[14] Quinn, P.J. (2010) *Prog. Lipid Res.*, *49*, 390-406
[15] Sonnino, S., and Prinetti, A. (2013) *Curr. Med. Chem.*, *20*, 4-21
[16] Brown, D.A. (2006) *Physiology*, *21*, 430-439
[17] Lingwood, D., and Simons, K. (2010) *Science*, *327*, 46-50
[18] Kraft, M.L. (2013) *Mol. Biol. Cell*, *24*, 2765-2768
[19] de Almeida, R.F.M., and Joly, E. (2013) *Front. Plant Sci.*,
[20] Dufourc, E. (2008) *J. Chem. Biol.*, *1*, 63-77

[21] Ole G.Mouritsen, Life As a Matter of Fat: The Emerging Science of Lipidomics, Springuer-Verlag, Berlin-Heidelberg-New York, 2005.

[22] Ipsen, J.H., Karlstrom, G., Mouritsen, O.G., Wennerstrom, H., Zuckermann, M.J. (1987) *Biochim. Biophys. Acta*, *905*, 162-172

[23] Marsh, D. (2010) *Biochim. Biophys. Acta -Biomembranes*, *1798*, 688-699

[24] Derek Marsh, Handbook of Lipid Bilayers, CRC Press, Taylor & Francis Group, 2013.

[25] Marsh, D. (2009) *Biochim. Biophys. Acta -Biomembranes*, *1788*, 2114-2123

[26] de Almeida, R.F.M., Loura, L.M.S., Prieto, M. (2009) *Chem. Phys. Lipids*, *157*, 61-77

[27] Davis, J.H., Clair, J.J., Juhasz, J. (2009) *Biophys. J.*, *96*, 521-539

[28] Dietrich, C., Bagatolli, L.A., Volovyk, Z.N., Thompson, N.L., Levi, M., Jacobson, K., Gratton, E. (2001) *Biophys. J.*, *80*, 1417-1428

[29] Edidin, M. (2003) *Annu. Rev. Bioph. Biom.*, *32*, 257-283

[30] de Almeida, R.F., Fedorov, A., Prieto, M. (2003) *Biophys. J.*, *85*, 2406-2416

[31] de Almeida, R.F.M., Loura, L.M.S., Fedorov, A., Prieto, M. (2005) *J. Mol. Biol.*, *346*, 1109-1120

[32] Hara, A., and Radin, N.S. (1978) *Anal.l Biochem.*, *90*, 420-426

[33] Barcelo-Coblijn, G., Martin, M.L., de Almeida, R.F.M., Noguera-Salva, M.A., Marcilla-Etxenike, A., Guardiola-Serrano, F., Luth, A., Kleuser, B., Halver, J.E., Escriba, P.V. (2011) *Proc. Natl. Acad. Sci. USA*, *108*, 19569-19574

[34] Silva, L.C., de Almeida, R.F., Castro, B.M., Fedorov, A., Prieto, M. (2007) *Biophys. J.*, *92*, 502-516

[35] Martin, M.L., Barcelo-Coblijn, G., de Almeida, R.F.M., Noguera-Salva, M.A., Teres, S., Higuera, M., Liebisch, G., Schmitz, G., Busquets, X., Escriba, P.V. (2013) *Biochim. Biophys. Acta -Biomembranes*, *1828*, 1405-1413

[36] Llado, V., Gutierrez, A., Martinez, J., Casas, J., Teres, S., Higuera, M., Galmes, A., Saus, C., Besalduch, J., Busquets, X., Escriba, P.V. (2010) *J. Cell. Mol. Med.*, *14*, 659-670

[37] Simons, K., and Vaz, W.L. (2004) *Annu. Rev. Biophys. Biomol. Struct.*, *33*, 269-295

[38] Bunge, A., Muller, P., Stockl, M., Herrmann, A., Huster, D. (2008) *Biophys. J.*, *94*, 2680-2690

[39] Gandhavadi, M., Allende, D., Vidal, A., Simon, S.A., McIntosh, T.J. (2002) *Biophys. J.*, *82*, 1469-1482
[40] Giocondi, M.C., Milhiet, P.E., Dosset, P., Le Grimellec, C. (2004) *Biophys. J.*, *86*, 861-869
[41] Yuan, C.B., Furlong, J., Burgos, P., Johnston, L.J. (2002) *Biophys. J.*, *82*, 2526-2535
[42] Marques, J.T., Viana, A.S., de Almeida, R.F. (2011) *Biochim. Biophys. Acta -Biomembranes*, *1808*, 405-414
[43] Vanegas, J.M., Faller, R., Longo, M.L. (2010) *Langmuir*, *26*, 10415-10418
[44] Vanegas, J.M., Contreras, M.F., Faller, R., Longo, M.L. (2012) *Biophys. J.*, *102*, 507-516
[45] Bretscher, M.S., and Munro, S. (1993) *Science*, *261*, 1280-1281
[46] Scolari, S., Engel, S., Krebs, N., Plazzo, A.P., de Almeida, R.F., Prieto, M., Veit, M., Herrmann, A. (2009) *J. Biol. Chem.*, *284*, 15708-15716
[47] Engel, S., Scolari, S., Thaa, B., Krebs, N., Korte, T., Herrmann, A., Veit, M. (2010) *Biochem. J.*, *425*, 567-573
[48] Filippov, A., Oradd, G., Lindblom, G. (2003) *Biophys. J.*, *84*, 3079-3086
[49] Filippov, A., Oradd, G., Lindblom, G. (2004) *Biophys. J.*, *86*, 891-896
[50] Elson E.L., Fried E., Dolbow J.E., Genin G.M. (2010) *Annu. Rev. Biophys.*, *39*, 207-226
[51] Shrivastava, S., and Chattopadhyay, A. (2007) *Biochem. Bioph. Res. Co.*, *356*, 705-710
[52] O'Shea, P. (2003) *Biochem. Soc. T.*, *31*, 990-996
[53] Starke-Peterkovic, T., Turner, N., Vitha, M.F., Waller, M.P., Hibbs, D.E., Clarke, R.J. (2006) *Biophys. J.*, *90*, 4060-4070
[54] Bastos, A.E.P., Marinho, H.S., Cordeiro, A.M., de Soure, A.M., de Almeida, R.F.M. (2012) *Chem. Phys. Lipids*, *165*, 577-588
[55] Pike, L.J. (2009) *J. Lipid Res.*, *50*, S323-S328
[56] Sezgin, E., Kaiser, H.J., Baumgart, T., Schwille, P., Simons, K., Levental, I. (2012) *Nat. Protoc.*, *7*, 1042-1051
[57] Baumgart, T., Hammond, A.T., Sengupta, P., Hess, S.T., Holowka, D.A., Baird, B.A., Webb, W.W. (2007) *Proc. Natl. Acad. Sci. USA*, *104*, 3165-3170
[58] Stockl, M., Plazzo, A.P., Korte, T., Herrmann, A. (2008) *J. Biol. Chem.*, *283*, 30828-30837
[59] Veatch, S.L., Cicuta, P., Sengupta, P., Honerkamp-Smith, A., Holowka, D., Baird, B. (2008) *Acs Chem. Biol.*, *3*, 287-293

[60] Meder, D., Moreno, M.J., Verkade, P., Vaz, W.L.C., Simons, K. (2006) *Proc. Natl. Acad. Sci. USA*, *103*, 329-334

[61] Sharma, P., Varma, R., Sarasij, R.C., Ira, Gousset, K., Krishnamoorthy, G., Rao, M., Mayor, S. (2004) *Cell*, *116*, 577-589

[62] Prior, I.A., Muncke, C., Parton, R.G., Hancock, J.F. (2003) *J. Cell Biol.*, *160*, 165-170

[63] de Almeida, R.F.M., Borst, J., Fedorov, A., Prieto, M., Visser, A.J.W.G. (2007) *Biophys. J.*, *93*, 539-553

[64] A.E.P. Bastos, S. Scolari, M. Stöckl, R.F.M. de Almeida, Chapter Three - Applications of Fluorescence Lifetime Spectroscopy and Imaging to Lipid Domains In Vivo, in: *Method. Enzymol.*, Imaging and Spectroscopic Analysis of Living Cells Optical and Spectroscopic Techniques, Academic Press, 2012, First Ed., pp. 57-81.

[65] Owen, D.M., Magenau, A., Williamson, D., Gaus, K. (2012) *BioEssays*, *34*, 739-747

[66] Eggeling, C., Ringemann, C., Medda, R., Schwarzmann, G., Sandhoff, K., Polyakova, S., Belov, V.N., Hein, B., von Middendorff, C., Schoenle, A., Hell, S.W. (2009) *Nature*, *457*, 1159-U121

[67] Henry, B., Moller, C., Dimanche-Boitrel, M.T., Gulbins, E., Becker, K.A. (2013) *Cancer Lett.*, *332*, 286-294

[68] Dumitru, C.A., Carpinteiro, A., Trarbach, T., Hengge, U.R., Gulbins, E. (2007) *Apoptosis*, *12*, 1533-1541

[69] Rebillard, A., Tekpli, X., Meurette, O., Sergent, O., LeMoigne-Muller, G., Vernhet, L., Gorria, M., Chevanne, M., Christmann, M., Kaina, B., Counillon, L., Gulbins, E., Lagadic-Gossmann, D., Dimanche-Boitrel, M.T. (2007) *Cancer Res.*, *67*, 7865-7874

[70] Grassme, H., Jekle, A., Riehle, A., Schwarz, H., Berger, J., Sandhoff, K., Kolesnick, R., Gulbins, E. (2001) *J. Biol. Chem.*, *276*, 20589-20596

[71] Grassme, H., Riethmuller, J., Gulbins, E. (2007) *Prog. Lipid Res.*, *46*, 161-170

[72] Wu, B.X., Clarke, C.J., Hannun, Y.A. (2010) *Neuromol. Med.*, *12*, 320-330

[73] Vitner, E.B., and Futerman, A.H. (2013) *Handb. Exp. Pharmacol.*, *2013/04/09*, 405-419

[74] Horres, C.R., and Hannun, Y.A. (2012) *Neurochem. Res.*, *37*, 1137-1149

[75] Bollinger, C.R., Teichgraber, V., Gulbins, E. (2005) *Biochim. Biophys. Acta*, *1746*, 284-294

[76] Carpinteiro, A., Dumitru, C., Schenck, M., Gulbins, E. (2008) *Cancer Lett.*, *264*, 1-10

[77] Zhang, Y., Li, X., Carpinteiro, A., Gulbins, E. (2008) *J. Immunol.*, *181*, 4247-4254

[78] Becker, K.A., Gellhaus, A., Winterhager, E., Gulbins, E. (2008) *Sub-Cell. Biochem.*, *49*, 523-538

[79] Stiban, J., Tidhar, R., Futerman, A.H. (2010) *Adv. Exp. Med. Biol.*, *688*, 60-71

[80] Pinto, S.N., Fernandes, F., Fedorov, A., Futerman, A.H., Silva, L.C., Prieto, M. (2013) *Biochim. Biophys. Acta*, *1828*, 2099-2110

[81] Pinto, S.N., Silva, L.C., de Almeida, R.F., Prieto, M. (2008) *Biophys. J.*, *95*, 2867-2879

[82] Pinto, S.N., Silva, L.C., Futerman, A.H., Prieto, M. (2011) *Biochim. Biophys. Acta*, *1808*, 2753-2760

[83] Castro, B.M., de Almeida, R.F., Silva, L.C., Fedorov, A., Prieto, M. (2007) *Biophys. J.*, *93*, 1639-1650

[84] Contreras, F.X., Basanez, G., Alonso, A., Herrmann, A., Goni, F.M. (2005) *Biophys. J.*, *88*, 348-359

[85] Busto, J.V., Fanani, M.L., De Tullio, L., Sot, J., Maggio, B., Goni, F.M., Alonso, A. (2009) *Biophys. J.*, *97*, 2717-2726

[86] Chiantia, S., Kahya, N., Schwille, P. (2007) *Langmuir*, *23*, 7659-7665

[87] De Tullio, L., Maggio, B., Hartel, S., Jara, J., Fanani, M.L. (2007) *Cell Biochem. Biophys.*, *47*, 169-177

[88] Fanani, M.L., Hartel, S., Oliveira, R.G., Maggio, B. (2002) *Biophys. J.*, *83*, 3416-3424

[89] Holopainen, J.M., Brockman, H.L., Brown, R.E., Kinnunen, P.K. (2001) *Biophys. J.*, *80*, 765-775

[90] Holopainen, J.M., Lehtonen, J.Y., Kinnunen, P.K. (1997) *Chem. Phys. Lipids*, *88*, 1-13

[91] Holopainen, J.M., Lemmich, J., Richter, F., Mouritsen, O.G., Rapp, G., Kinnunen, P.K. (2000) *Biophys. J.*, *78*, 2459-2469

[92] Busto, J.V., Sot, J., Requejo-Isidro, J., Goni, F.M., Alonso, A. (2010) *Biophys. J.*, *99*, 1119-1128

[93] Sot, J., Aranda, F.J., Collado, M.I., Goni, F.M., Alonso, A. (2005) *Biophys. J.*, *88*, 3368-3380

[94] Sot, J., Goni, F.M., Alonso, A. (2005) *Biochim. Biophys. Acta*, *1711*, 12-19

[95] Veiga, M.P., Arrondo, J.L., Go+¦i, F.M., Alonso, A. (1999) *Biophys. J.*, *76*, 342-350

[96] Castro, B.M., Silva, L.C., Fedorov, A., de Almeida, R.F., Prieto, M. (2009) *J. Biol. Chem.*, *284*, 22978-22987

[97] Silva, L., de Almeida, R.F., Fedorov, A., Matos, A.P., Prieto, M. (2006) *Mol. Membr. Biol.*, *23*, 137-148
[98] Silva, L.C., Futerman, A.H., Prieto, M. (2009) *Biophys. J.*, *96*, 3210-3222
[99] Shah, J., Atienza, J.M., Duclos, J., Rawlings, A.V., Dong, Z., Shipley, G.G. (1995) *J. Lipid Res.*, *36*, 1936-1944
[100] Dupuy, F., Fanani, M.L., Maggio, B. (2011) *Langmuir*, *27*, 3783-3791
[101] Maggio, B. (2004) *Chem. Phys. Lipids*, *132*, 209-224
[102] De Tullio, L., Maggio, B., Fanani, M.L. (2008) *J. Lipid Res.*, *49*, 2347-2355
[103] Catapano, E.R., Arriaga, L.R., Espinosa, G., Monroy, F., Langevin, D., Lopez-Montero, I. (2011) *Biophys. J.*, *101*, 2721-2730
[104] Espinosa, G., Lopez-Montero, I., Monroy, F., Langevin, D. (2011) *Proc. Natl. Acad. Sci. USA*, *108*, 6008-6013
[105] Huang, H.W., Goldberg, E.M., Zidovetzki, R. (1996) *Biochem. Bioph. Res. Co.*, *220*, 834-838
[106] Huang, H.W., Goldberg, E.M., Zidovetzki, R. (1998) *Eur. Biophys. J.*, *27*, 361-366
[107] Hsueh, Y.W., Giles, R., Kitson, N., Thewalt, J. (2002) *Biophys. J.*, *82*, 3089-3095
[108] Holopainen, J.M., Subramanian, M., Kinnunen, P.K. (1998) *Biochemistry*, *37*, 17562-17570
[109] Massey, J.B. (2001) *Biochim. Biophys. Acta*, *1510*, 167-184
[110] Carrer, D.C., and Maggio, B. (1999) *J. Lipid Res.*, *40*, 1978-1989
[111] Fidorra, M., Duelund, L., Leidy, C., Simonsen, A.C., Bagatolli, L.A. (2006) *Biophys. J.*, *90*, 4437-4451
[112] Sot, J., Bagatolli, L.A., Goni, F.M., Alonso, A. (2006) *Biophys. J.*, *90*, 903-914
[113] Holopainen, J.M., Angelova, M.I., Kinnunen, P.K. (2000) *Biophys. J.*, *78*, 830-838
[114] Boulgaropoulos, B., Arsov, Z., Laggner, P., Pabst, G. (2011) *Biophys. J.*, *100*, 2160-2168
[115] Megha, and London, E. (2004) *J. Biol. Chem.*, *279*, 9997-10004
[116] Alanko, S.M., Halling, K.K., Maunula, S., Slotte, J.P., Ramstedt, B. (2005) *Biochim. Biophys. Acta*, *1715*, 111-121
[117] Nybond, S., Bjorkqvist, Y.J., Ramstedt, B., Slotte, J.P. (2005) *Biochim. Biophys. Acta*, *1718*, 61-66
[118] Sot, J., Ibarguren, M., Busto, J.V., Montes, L.R., Goni, F.M., Alonso, A. (2008) *FEBS Lett.*, *582*, 3230-3236

[119] Megha, Sawatzki, P., Kolter, T., Bittman, R., London, E. (2007) *Biochim. Biophys. Acta*, *1768*, 2205-2212

[120] Megha, Bakht, O., London, E. (2006) *J. Biol. Chem.*, *281*, 21903-21913

[121] Scheffer, L., Fargion, I., Addadi, L. (2006) *Chembiochem*, *7*, 1680-1682

[122] Scheffer, L., Solomonov, I., Weygand, M.J., Kjaer, K., Leiserowitz, L., Addadi, L. (2005) *Biophys. J.*, *88*, 3381-3391

[123] Goldschmidt-Arzi, M., Shimoni, E., Sabanay, H., Futerman, A.H., Addadi, L. (2011) *J. Struct. Biol.*, *175*, 21-30

[124] Scheffer, L., Futerman, A.H., Addadi, L. (2007) *Chembiochem.*, *8*, 2286-2294

[125] N. Ferreira, M. Goldschmidt-Arzi, H. Sabanay, J. Storch, T. Levade, M. Ribeiro, L. Addadi, A. Futerman, Accumulation of Ordered Ceramide-Cholesterol Domains in Farber Disease Fibroblasts, *JIMD Reports*, Springer Berlin Heidelberg, 2013, pp. 1-7.

[126] Chiantia, S., Kahya, N., Ries, J., Schwille, P. (2006) *Biophys. J.*, *90*, 4500-4508

[127] Johnston, I., and Johnston, L.J. (2006) *Langmuir.*, *22*, 11284-11289

[128] Staneva, G., Momchilova, A., Wolf, C., Quinn, P.J., Koumanov, K. (2009) *Biochim. Biophys. Acta*, *1788*, 666-675

[129] Boulgaropoulos, B., Rappolt, M., Sartori, B., Amenitsch, H., Pabst, G. (2012) *Biophys. J.*, *102*, 2031-2038

[130] Staneva, G., Chachaty, C., Wolf, C., Koumanov, K., Quinn, P.J. (2008) *Biochim. Biophys. Acta*, *1778*, 2727-2739

[131] Popov, J., Vobornik, D., Coban, O., Keating, E., Miller, D., Francis, J., Petersen, N.O., Johnston, L.J. (2008) *Langmuir*, *24*, 13502-13508

[132] Sullan, R.M., Li, J.K., Zou, S. (2009) *Langmuir*, *25*, 7471-7477

[133] Ira, and Johnston, L.J. (2008) *Biochim. Biophys. Acta*, *1778*, 185-197

[134] Ira, Zou, S., Ramirez, D.M., Vanderlip, S., Ogilvie, W., Jakubek, Z.J., Johnston, L.J. (2009) *J. Struct.l Biol.*, *168*, 78-89

[135] Taniguchi, Y., Ohba, T., Miyata, H., Ohki, K. (2006) *Biochim. Biophys. Acta*, *1758*, 145-153

[136] Ruiz-Arguello, M.B., Veiga, M.P., Arrondo, J.L., Goni, F.M., Alonso, A. (2002) *Chem.. Phys. Lipids*, *114*, 11-20

[137] Jungner, M., Ohvo, H., Slotte, J.P. (1997) *Biochim. Biophys. Acta*, *1344*, 230-240

[138] Ruiz-Arguello, M.B., Basanez, G., Goni, F.M., Alonso, A. (1996) *J. Biol. Chem.*, *271*, 26616-26621

[139] Contreras, F.X., Villar, A.V., Alonso, A., Kolesnick, R.N., Goni, F.M. (2003) *J. Biol. Chem.*, *278*, 37169-37174

[140] Hartel, S., Fanani, M.L., Maggio, B. (2005) *Biophys. J.*, *88*, 287-304
[141] Ruiz-Arguello, M.B., Goni, F.M., Alonso, A. (1998) *J. Biol. Chem.*, *273*, 22977-22982
[142] Carrer, D.C., Kummer, E., Chwastek, G., Chiantia, S., Schwille, P. (2009) *Soft Matter*, *5*, 3279-3286
[143] Chiantia, S., Ries, J., Chwastek, G., Carrer, D., Li, Z., Bittman, R., Schwille, P. (2008) *Biochim. Biophys. Acta*, *1778*, 1356-1364
[144] Castro, B.M., de Almeida, R.F., Goormaghtigh, E., Fedorov, A., Prieto, M. (2011) *Biophys. J.*, *101*, 1632-1641
[145] Huang, H.W., Goldberg, E.M., Zidovetzki, R. (1999) *Biophys. J.*, *77*, 1489-1497
[146] Fanani, M.L., and Maggio, B. (1998) *Lipids*, *33*, 1079-1087
[147] Pike, L.J. (2006) *J. Lipid Res.*, *47*, 1597-1598
[148] Grassme, H., Cremesti, A., Kolesnick, R., Gulbins, E. (2003) *Oncogene*, *22*, 5457-5470
[149] Cubi, R., Candalija, A., Ortega, A., Gil, C., Aguilera, J. (2013) *PloS One*, *8*, e68055
[150] Grassme, H., Jendrossek, V., Bock, J., Riehle, A., Gulbins, E. (2002) *J. Immunol.*, *168*, 298-307
[151] Cremesti, A., Paris, F., Grassme, H., Holler, N., Tschopp, J., Fuks, Z., Gulbins, E., Kolesnick, R. (2001) *J. Biol. Chem.*, *276*, 23954-23961
[152] Paris, F., Grassme, H., Cremesti, A., Zager, J., Fong, Y., Haimovitz-Friedman, A., Fuks, Z., Gulbins, E., Kolesnick, R. (2001) *J. Biol. Chem.*, *276*, 8297-8305
[153] Abdel Shakor, A.B., Kwiatkowska, K., Sobota, A. (2004) *J. Biol. Chem.*, *279*, 36778-36787
[154] Korzeniowski, M., Shakor, A.B., Makowska, A., Drzewiecka, A., Bielawska, A., Kwiatkowska, K., Sobota, A. (2007) *Cell. Physiol. Biochem.*, *20*, 347-356
[155] Avota, E., Gulbins, E., Schneider-Schaulies, S. (2011) *PLoS Pathog.*, *7*, e1001290
[156] Trajkovic, K., Hsu, C., Chiantia, S., Rajendran, L., Wenzel, D., Wieland, F., Schwille, P., Brugger, B., Simons, M. (2008) *Science*, *319*, 1244-1247
[157] A. Draeger and E. Babiychuk, Ceramide in Plasma Membrane Repair, in: E.Gulbins and I.Petrache (Eds.), Sphingolipids in Disease, Handbook of Experimental Pharmacology, Springer Vienna, 2013, pp. 341-353.

[158] H. Grassme and K. Becker, Bacterial Infections and Ceramide, in: E.Gulbins and I.Petrache (Eds.), Sphingolipids in Disease, Handbook of Experimental Pharmacology, Springer Vienna, 2013, pp. 305-320.

[159] Dreschers, S., Franz, P., Dumitru, C., Wilker, B., Jahnke, K., Gulbins, E. (2007) *Cell. Physiol. Biochem.*, *20*, 241-254

[160] Bezombes, C., Grazide, S., Garret, C., Fabre, C., Quillet-Mary, A., Muller, S., Jaffrezou, J.P., Laurent, G. (2004) *Blood*, *104*, 1166-1173

[161] Grether-Beck, S., Salahshour-Fard, M., Timmer, A., Brenden, H., Felsner, I., Walli, R., Fullekrug, J., Krutmann, J. (2008) *Oncogene*, *27*, 4768-4778

[162] Lacour, S., Hammann, A., Grazide, S., Lagadic-Gossmann, D., Athias, A., Sergent, O., Laurent, G., Gambert, P., Solary, E., Dimanche-Boitrel, M.T. (2004) *Cancer Res.*, *64*, 3593-3598

[163] Stancevic, B., and Kolesnick, R. (2010) *FEBS Lett.*, *584*, 1728-1740

[164] Pinto, S.N., Laviad, E.L., Stiban, J., Kelly, S.L., Merrill, A.H., Prieto, M., Futerman, A.H., Silva, L.C. (2014) *J. Lipid Res.*, doi:10.1194/jlr.M042002

In: Lipid Rafts
Editor: Dan Sillence

ISBN: 978-1-62948-961-2

Chapter III

Caveolae as Specialized Lipid Rafts: Diverse Roles in the Regulation of Potassium Channel Signalling

Dominic Byrne and Caroline Dart
Institute of Integrative Biology, University of Liverpool,
Liverpool, UK

Abstract

Caveolae represent a distinct subtype of lipid raft. They appear as small vesicular pits (50–100 nm in diameter) on the surface of most cells and form when discrete regions of the plasma membrane rich in cholesterol and sphingolipids associate with the proteins cavin and caveolin. The precise physiological role of these structures is elusive and somewhat controversial, but they have been implicated in vesicular trafficking, as sites of cholesterol influx/efflux and in detecting and transducing membrane stretch. Biochemical purification and analysis of caveolae also suggests that these membrane compartments are populated by a wealth of signalling proteins, including heterotrimeric G-proteins, the small GTP-binding protein Ras, receptor/non-receptor tyrosine kinases, serine/threonine kinases, various classes of G-protein-coupled receptors and both voltage- and ligand-gated ion channels. Aside from the obvious implications to signalling speed and efficiency of spatially confining interacting signalling molecules, caveolae offer a unique lipid environment that may also impact on signalling events. The presence of tightly-packed aggregates of cholesterol and sphingolipids within caveolae alters membrane thickness and rigidity, influencing the function of proteins that undergo significant conformational changes during activation/inactivation such as ion channels and receptors. Direct interaction between raft-enriched lipids like cholesterol and sphingomyelin also modulates protein activity as does association with caveolae-specific scaffold proteins such as caveolin. Potassium channels represent a particularly good example of a set of proteins that undergo diverse modes of regulation through compartmentalization within caveolae. This short communication discusses recent research that highlights the role of these specialized lipid rafts in controlling potassium channel kinetics, signalling efficiency, trafficking, and surface expression and emphasises potential physiological roles for caveolae in signalling events driven by potassium channel activity such as insulin secretion, the regulation of vascular tone and cardioprotection.

Introduction

Caveolae represent a distinct subtype of lipid raft. They appear as small vesicular pits (60–80 nm in diameter) on the surface of most cells and form when discrete regions of the plasma membrane rich in cholesterol and sphingolipids associate with the proteins cavin and caveolin (reviewed by [1]). They have been known about since the 1950's, when their distinct morphology caught the eye of the early electron microscopists [2, 3], yet remain one of the few organelles left within cells without a clearly defined function. Over the years a number of diverse physiological roles have been assigned to caveolae including vesicular trafficking, as sites of cholesterol influx/efflux and in detecting and transducing membrane stretch [1, 4, 5]. Biochemical purification and analysis of caveolae also suggests that these membrane compartments are populated by a wealth of signalling proteins, including heterotrimeric G-proteins, the small GTP-binding protein Ras, receptor/non-receptor tyrosine kinases, serine/threonine kinases, various classes of G-protein-coupled receptors and both voltage- and ligand-gated ion channels [5-7]. This apparent ability to recruit signalling molecules has led to the idea that caveolae act to functionally compartmentalise cell-surface signalling. Aside from the obvious implications to signalling speed and efficiency of spatially confining interacting signalling molecules, caveolae offer a unique lipid environment that may also impact on signalling events. The physical characteristics of the membrane within caveolae are distinct from that of the bulk bilayer due to the presence of tightly-packed aggregates of cholesterol and sphingolipids, which alter membrane thickness and rigidity [8-10]. Signalling proteins that undergo significant conformational changes during activation/inactivation, such as the gating motions of ion channels can be affected by this change in surrounding membrane composition [8]. Direct interaction with raft-enriched lipids is also known to modulate protein activity as is association with caveolae-specific scaffold proteins such as caveolin [11]. Potassium ion channels represent a particularly good example of a set of proteins that undergo diverse modes of regulation through their association with caveolae. K^+ channels are a large family of integral membrane proteins that form aqueous pores through which K^+ can flow. They are unique amongst ion channels in that they are found in virtually all types of cells in all organisms where they perform a range of biological functions. The opening of K^+ channels, which occurs in response to a range of different signals, leads almost universally to the efflux of K^+ from the cell causing the membrane potential to become more negative. This most famously results in the repolarization of the membrane during an action potential, but also modulates fundamental processes like muscle contraction or hormone-secretion by switching off voltage-sensitive Ca^{2+} channels that allow the influx of Ca^{2+} into cells. This brief commentary discusses some of the recent research that highlights the role of caveolae in controlling K^+ channel kinetic activity, signalling efficiency, trafficking, and surface expression (*Figure 1*), and emphasises potential physiological roles for caveolae in the regulation of insulin secretion, of vascular tone and in cardioprotection.

Regulation of K^+ Channel Signalling through Spatial Compartmentation

Biochemical and proteomic analysis of isolated caveolae suggest that signalling molecules make up a large proportion of the protein complement of these membrane compartments [6]. This has led to the idea that caveolae act as hubs that spatially and functionally organise signalling pathways [4, 5, 7]. There are several examples, particularly in the heart, where caveolae appear to either aggregate or segregate receptor-driven pathways that modulate the activity of different subsets of K^+ channel [12-15]. Adenosine A1 receptor-mediated activation of cardiac ATP-sensitive K^+ (K_{ATP}) channels is, for example, largely eliminated by disrupting caveolae with the cholesterol-depleting agent methyl-β-cyclodextrin, or by use of small interfering RNA (siRNA) to knock down the muscle-specific caveolin isoform, caveolin-3 [12]. The activation of cardiac K_{ATP} channels by adenosine, which is released from metabolically-compromised cells, causes a pronounced shortening of the cardiac action potential, a reduction in Ca^{2+} entry during the plateau phase and ultimately contractile failure. This is believed to conserve ATP and thus minimize cell damage during metabolic stress [16, 17]. Compartmentation of adenosine receptors and K_{ATP} channels in caveolae may thus be essential for channel regulation during periods of cardiac ischaemia [12]. The spatial organisation of signalling pathways also regulates the activity of arterial K_{ATP} channels which play a critical role in the regulation of vascular tone. For example, efficient K_{ATP} channel activation by cAMP-dependent protein kinase A requires co-localisation with adenylyl cyclase in the same smooth muscle caveolar compartment [24].

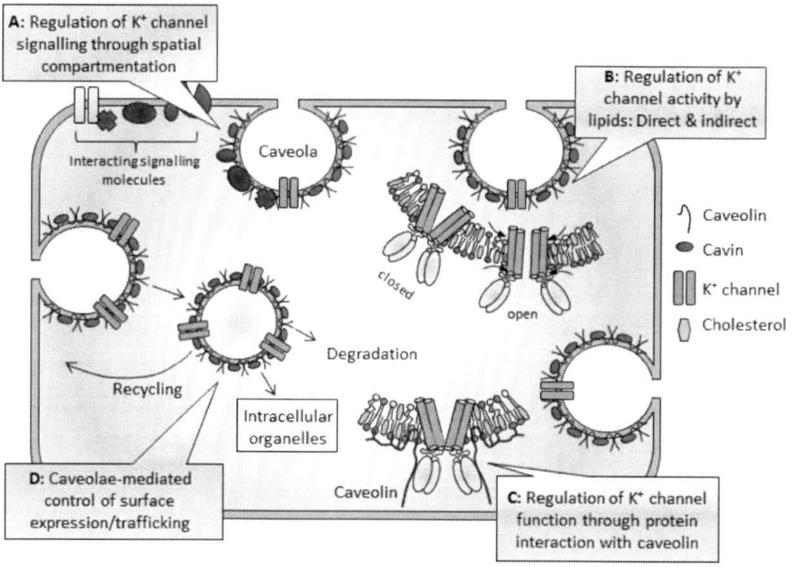

Figure 1. Schematic showing the diverse roles of caveolae in controlling K^+ channel function. Caveolae have the ability to either aggregate or segregate receptor-driven pathways that modulate the activity of different subsets of K^+ channel (A); modulate channel activity through the local lipid environment, either directly through interactions with cholesterol or sphingolipids, or indirectly through changes in general membrane properties (B); regulate channel activity through direct or indirect interactions with caveolae-resident proteins such as caveolin (C); control channel surface density by regulating trafficking (D).

Slowing of the heart rate following parasympathetic activation of cardiac M2 muscarinic receptors involves the activation of G-protein-gated inward rectifier K^+ channels (GIRK) channels. This occurs via the direct binding of $G_{\beta\gamma}$ subunits, which are released from inhibitory G proteins ($G_{i/o}$) following M2 muscarinic receptor activation, to GIRK channel subunits (Kir3.1/3.4). G_s-coupled receptors, such as the β-adrenergic receptor, do not activate this current despite also stimulating $G_{\beta\gamma}$ subunit release. A possible explanation for this is that while $G_{i\alpha(1/2)}$ subunits were fairly evenly distributed between caveolar and non-caveolar regions of the membrane, stimulatory G protein subunits are found exclusively in caveolae, tucked away from easy access to Kir3.1/3.4 channel subunits and M2 muscarinic receptors which localize predominantly to the non-caveolar bulk of the membrane [13]. Cardiac GIRK channels can also be inhibited by the depletion of membrane phosphatidylinositol 4,5-bisphosphate $PI(4,5)P_2$ that occurs following activation of G_q-coupled receptors by neurotransmitters and hormones. However, endothelin-1, but not bradykinin, inhibits GIRK activity even though both induce $PI(4,5)P_2$ hydrolysis in cardiomyocytes [18]. Biochemical analysis shows that agonist stimulation of mouse atrial myocytes induces a transient co-localization of GIRK channels subunits and endothelin receptors in caveolae, while bradykinin receptors are excluded [14]. Treatment of these cells with latrunculin B to remove any cytoskeletal restriction on $PI(4,5)P_2$ diffusion abolishes GIRK regulation by G_q-coupled receptors without affecting the transient targeting to caveolae. These data suggest that low $PI(4,5)P_2$ mobility and confinement of the lipid to caveolae is important and that selective recruitment of receptors and effectors to these microdomains ensures specificity [14].

Regulation of K^+ Channel Function through Direct Interaction of Proteins with Caveolin

The activity of K^+ channels within caveolae may also be influenced through interaction with caveolae-resident proteins. Caveolins are a family of cholesterol-binding membrane proteins (caveolin-1, -2 and -3) that coat the intracellular surface of caveolae. Aside from roles in caveolae formation and stability, caveolins interact with many caveolae-localized signalling molecules including heterotrimeric G proteins, Src family tyrosine kinases, H-Ras, endothelial nitric oxide (eNOS) and a number of ion channels via a ~ 20 amino acid N-terminal region on the caveolin molecule known as the caveolin scaffolding domain (CSD) [5, 7]. In most cases interaction with caveolin appears to hold proteins in an inactive conformation (although some proteins, such as the insulin receptor, are positively regulated by caveolin [19]) while enrichment in caveolae ensures a rapid response upon stimulation. This 'caveolin signalling hypothesis' is supported to some degree by the fact that loss of caveolins disinhibits a number of aspects of cell signalling, particularly proliferative and nitric oxide signalling pathways [20]. It should however be borne in mind that loss of caveolin may also cause loss of caveolae and compartmentation. Caveolin-protein interactions are commonly considered to be mediated via the CSD and an aromatic-rich caveolin binding motif (CBM) on partner proteins (ΦXΦXXXXΦ, ΦXXXXΦXXΦ or ΦXΦXXXXΦXXΦ, where Φ is an aromatic amino acid and X is an unspecified amino acid) [21]. The CBM resembles a typical linear motif – a short, simple sequence independently evolved many times in different proteins for a specific function. However, bioinformatic and structure-based analysis has recently shown that the canonical CBM is unlikely to be a common characteristic of caveolin-target interactions and that interfaces between caveolin and target proteins are more structurally diverse than presently appreciated [22, 23].

The vasculature isoform of the K_{ATP} channel has been shown to localize to smooth muscle caveolae and interact with the CSD of caveolin-1 [24 -26]. These channels have been described in smooth muscle cells from a variety of arterial beds where they provide a background K^+ conductance important in the regulation of membrane potential and thus smooth muscle contractility and blood flow [27-30]. Recent work on recombinant and native vascular K_{ATP} channels suggests that interaction with the scaffolding domain of caveolin-1 has an inhibitory effect on channel activity by altering the channel's sensitivity to its physiological regulator MgADP [31]. Nucleotide regulation of K_{ATP} channels in the presence of Mg^{2+} involves binding of ATP or ADP to two large intracellular loops on the accessory sulphonylurea receptor (SUR) subunits that contain consensus sequences for nucleotide binding and hydrolysis [32-34]. The CSD of caveolin may thus target residues on the SUR to affect channel activity.

The activity of recombinant cardiac K_{ATP} channels in HEK293T cells is suppressed by coexpression with caveolin-3, but not caveolin-1, suggesting some specificity in regulation [12, 35]. In line with these findings, cardiac K_{ATP} channels subunits (Kir6.2/SUR2A) coimmunoprecipitate with caveolin-3 but not caveolin-1.

In the pancreatic βTC-6 cell line caveolin-1 coimmunoprecipitates with the K_{ATP} channel pore-forming subunit, Kir6.2, indicating that there may also be a physical interaction between these two proteins [36]. Pancreatic K_{ATP} channels help maintain the β cell resting membrane potential and are involved in glucose-dependent insulin secretion [37]. Elevation of blood glucose results in increased glucose uptake and its subsequent metabolism by β cells, leading to a rise in the intracellular levels of ATP and a fall of ADP. This closes active K_{ATP} channels resulting in a reduction of K^+ efflux and induces a membrane depolarization which activates L-type voltage-dependent Ca^{2+} channels, increasing Ca^{2+} influx and triggering the Ca^{2+}-dependent secretion of insulin. Clinically, sulphonylurea drugs cause a reduction in blood glucose by inhibiting pancreatic K_{ATP} channels and thereby triggering insulin secretion in the same chain of events as glucose. Functionally, siRNA knock-down of caveolin-1 significantly reduces both glucose and sulphonylurea-stimulated insulin secretion [36]. Insulin secretion stimulated by direct depolarization of the membrane, therefore bypassing K_{ATP} channels, is unaffected by caveolin-1 knockdown, suggesting a specific functional link between caveolin and pancreatic K_{ATP} channels. The mechanism behind caveolin-induced channel regulation is unclear although a reduction in caveolin-1 expression in the pancreatic β cell line MIN6 increases the activity of Cdc42, a small GTPase that upregulates $PI(4,5)P_2$ synthesis [38]. Caveolin-1 contains a motif within its scaffolding domain that is conserved in guanine nucleotide dissociation inhibitors (GDIs), which suggests that caveolin-1 may behave as a GDI in β cells, binding to and maintaining Cdc42 in an inactive state. It is thus possible that increased levels of the channel activator $PI(4,5)P_2$ following Cdc42 activation may antagonise the glucose-induced inhibitory changes in ATP/ADP, preventing channel closure and the initiation of insulin secretion.

Other K^+ channels that have been shown to be regulated by caveolin include the hyperpolarization-activated cardiac pacemaker channel, HCN4, where caveolin-3 regulates current density and activation/inactivation kinetics [39], and the channel encoded by the human ether-a-go-go-related gene (*hERG*), which forms the rapidly activating delayed rectifier K^+ channel (*I*Kr) involved in cardiac repolarization. Yeast 2-hybrid screens of a human heart library with the cytosolic (N and C terminal) regions of the *hERG* channel have identified caveolin-1 as a potential partner [40]. Interaction with caveolin negatively regulates *hERG* channel activity, significantly reducing current amplitude and slowing the deactivation of tail currents. It is as yet unclear whether it is the interaction with caveolin *per se* that regulates channel activity or whether caveolin acts to cluster the channel with kinases and/or phosphatases that regulate its behaviour.

Caveolae-Mediated Control of Surface Expression/Trafficking

High-resolution 3-dimensional electron microscopic analysis has revealed that caveolae are interconnected by a dense filamentous network formed by the actin cytoskeleton and microtubules [41]. This is consistent with the cytoskeleton playing a role in the cellular distribution of caveolae and may also indicate that caveolae have the ability to undergo dynamic trafficking as a means of controlling their surface density, the surface expression of resident proteins, or as a means of relaying surface signals to intracellular organelles.

Surface expression of the *hERG* channel has been shown to be regulated by caveolin-3-mediated endocytic degradation. It is likely that this involves caveolin-3 recruiting the ubiquitin ligase, Nedd4-2, from the cytosol to the membrane to enhance channel ubiquitination and subsequent caveolin-mediated internalization [42]. Caveolin-1 has additionally been shown to act as a protein chaperone necessary for the surface targeting of $K_V1.5$, and naturally occurring mutants of caveolin-1 which accumulate in the Golgi complex and fail to reach the surface membrane also trap Kv1.5 intracellularly and prevent functional channel expression at the cell surface [43]. Similarly in cell lines, caveolin-1 has been suggested to play a dual role in regulating surface targeting of the large conductance Ca^{2+}-activated (BK) channel α subunit, Slo1, by maintaining channels in intracellular compartments and/or anchoring channels in the membrane [44]. In myometrial smooth muscle cells, BK_{Ca} channels are localized to caveolae as part of a complex that includes caveolin and α- and γ-actin [45]. In the same study immunoelectron microscopy revealed the presence of BK_{Ca} channels in unfused caveolae just beneath the membrane. This raises the possibility that the addition or removal of BK_{Ca} channels to the cell surface via caveolae delivery and removal could act as a dynamic means of buffering K^+ channel numbers and thus cellular excitability. A similar mechanism has been proposed by Jiao *et al* for the regulation of vascular tone whereby the vasoconstrictor angiotensin II acts via PKC to facilitate caveolin-1- and dynamin-dependent internalization of vascular K_{ATP} channels [26]. The activity of these channels contributes a vasodilating drive to vascular tone and thus their removal from the membrane promotes vasoconstriction. Ca^{2+}-dependent caveolar trafficking has also been shown to alter the trafficking and surface expression of small conductance Ca^{2+}-activated (SK_{Ca}) channels in aortic endothelial cells [46]. Thus SK_{Ca} channel activity in this tissue, which is important in modulating endothelial cell membrane potential and the membrane potential and contractile state of the underlying smooth muscle, seems to be regulated not only by changes in $[Ca^{2+}]_i$, to directly activate the channel, but also by dynamic surface expression via Ca^{2+}-dependent trafficking which may act to fine-tune the contractile response.

Mutations of the muscle-specific caveolin isoform, caveolin-3, have been associated with Long QT Syndrome 9 (LQT9), a phenotype characterized by a prolongation of the cardiac action potential in individuals with otherwise structurally normal hearts [47]. One of the abnormalities seen in some patients suffering from LQT9 is the appearance of prominent U waves on electrocardiograms [47]. This is suggestive of loss of function of cardiac inwardly rectifying K^+ channels, Kir2.1, which mediate the *I*K1 current that regulates the late phase of cardiac action potential repolarization and stabilizes the resting membrane potential. Normal Kir2.1 trafficking to the surface membrane is disrupted by mutations of caveolin-3 clinically associated with LQT9, resulting in reduced Kir2.1 surface expression and thus current density [48]. Loss of Kir2.1 function may thus contribute to the mechanism of arrhythmia generation in caveolin 3-mediated LQT9.

Caveolae have also been implicated in the transmission of signals from the cell surface to intracellular organelles. Ischaemic preconditioning is a phenomenon whereby brief intermittent periods of ischaemia in the heart improve the recovery of contractile function and reduce infarct size following a subsequent prolonged ischaemic event brought about by the complete block of blood flow to the heart muscle. Ischaemic preconditioning can be reduced by the sulphonylurea glibenclamide and mimicked by K_{ATP} channel openers such as bimakalim and diazoxide, suggesting a possible role for K_{ATP} channels (reviewed by [49]). Though controversial (for discussion see [40]), one possible mechanism for cardioprotection arises via the transmission of a signal from cell surface receptors through the cytosol to the mitochondrial outer membrane where it initiates intramitochondrial signalling by opening K_{ATP} channels on the inner mitochondrial membrane [51]. The opening of mitochondrial K_{ATP} channels causes an increase in the production of reactive oxygen species (ROS) which leads to the inhibition of the mitochondrial permeability transition (MPT), a high conductance pore in the inner mitochondrial membrane that is the main cause of necrotic cell death in ischaemic reperfusion injury [52]. One hypothesis suggests that the transmission of the signal from the cell surface to the outer membrane of the mitochondria occurs via vesicular multimolecular signalling complexes termed signalosomes that originate in caveolae [51]. According to this hypothesis, occupied bradykinin receptors migrate to caveolae where they are assembled into complexes with interacting signalling molecules. The whole caveolae containing the multimolecular complex then pinches off into the cell and travels as an intact vesicle to the mitochondria. Functionally active signalosomes containing a range of signalling molecules including caveolin-3, eNOS, guanylyl cyclase, cGMP-dependent protein kinase, have been recovered from mitochondrial suspensions from bradykinin or ischaemic preconditioned hearts [53]. These isolated signalosomes can be dissolved by cholesterol-depletion using methyl-β-cyclodextrin, but are resistant to solubilisation in cold non-ionic detergent suggesting their origin as caveolae.

Regulation of K^+ Channel Activity by Lipids

Ion channels can be both directly and indirectly influenced by the lipid composition of the membrane in which they are embedded. Caveolar and non-caveolar rafts form from tightly-packed aggregates of cholesterol and sphingolipids and this lipid composition influences physical characteristics of the membrane such as thickness, curvature and the ability to bend and compress. The conformational changes that occur when a channel protein opens or closes cause a local disturbance in the surrounding bilayer. The overall energetic cost of a channel transition will thus include not only the intrinsic channel activation energy, but also energy associated with membrane deformation and the energetics and kinetics of channel gating will therefore be modulated by the local lipid environment (reviewed by [8]). There is also strong evidence for direct modulation of K^+ channel activity through interactions with surrounding lipids. One of the best characterised examples is the effect of cholesterol on inwardly rectifying K^+ (Kir) channels. Kir2 channel activity is strongly suppressed by an increase in membrane cholesterol. This is independent of changes in general membrane properties since epicholesterol, an optical isomer of cholesterol that differs only in the rotational angle of the hydroxyl group at position 3 and essentially mimics the effects of cholesterol on membrane properties, is unable modulate Kir channel activity [54]. Cholesterol has no effect on the unitary conductance and little effect on the open probability of the channels [55]. This suggests that cholesterol affects functional channel numbers, possibly by stabilizing the closed state of the channel leading to a population of silent and inactive channels in the cell membrane. Kir channels lack the known cholesterol recognition/interaction amino acid consensus (CRAC) motif (–L/V-(X)(1-5)-Y-(X)(1-5)-R/K, where (X)(1-5) represents between 1 and 5 amino acids of any type [56]) and the actual binding site remains elusive. However, recent studies have identified a circular belt of cholesterol-sensitive residues around the cytosolic pore of the channel close to its interface with the membrane [57]. There is also a partial overlap between residues important in cholesterol sensitivity and residues involved with $PI(4,5)P_2$ channel activation. It seems likely that conformational changes induced by cholesterol or $PI(4,5)P_2$ to stabilize the closed or open conformations respectively share a common pathway through specific regions of the cytosolic pore at the membrane interface. Other K^+ channels that show cholesterol sensitivity have known cholesterol interaction sites. The BK_{Ca} channel α subunit, Slo1 has 7 CRAC motifs (CRAC4-10) in its long cytosolic C tail domain. CRAC4, which is adjacent to the membrane, primarily confers cholesterol sensitivity, although successive deletion of CRAC5-10 reduces cholesterol sensitivity suggesting they also play a role [58].

Interaction between channel proteins and sphingolipids may also regulate voltage-sensing and/or channel gating. Treatment of cells with sphingomyelinase, an enzyme that cleaves the positively charged choline group from sphingomyelin, results in a pronounced negative shift in the activation of the voltage-gated K^+ channel, $K_v2.1$ and has led to the suggestion that key channel residues may interact with lipid head groups to modify channel activity [59]. Removal of the choline headgroup from sphingomyelin generates ceramide, which modulates $K_v1.3$ and *h*ERG channel activity, although the potential mechanism/s is unclear [60, 61].

Conclusion

Caveolae remain enigmatic cellular structures that resist easy functional classification. Their role in the regulation of the functional activity of K^+ channels ranges from direct effects on channel kinetics through control of trafficking and surface expression, to the spatial (and perhaps temporal) organisation of signalling partners. This complex and multi-layered regulatory capacity will clearly extend to other classes of protein. Their potential importance in normal cellular activity is reinforced by the findings that mutations in proteins essential for caveolae formation, caveolins and cavins, are linked to the development of a number of diseases in humans, including muscular dystrophies, lipodystrophy, long QT syndromes and several forms of cancer [47, 62-64]. Recent studies also show up-regulation of caveolin-1 in the endothelium and vascular smooth muscle of two rat models of chronic hypertension [65] and in the pulmonary arterial media of lungs obtained from patients with idiopathic pulmonary arterial hypertension (IPAH) [66]. It is as yet unclear whether these changes are causal or an effect of the disease, but interestingly, there is emerging evidence that regions of the caveolin molecule can be used to reverse aberrant signalling brought about by changes in caveolin expression. Cardiac ischaemic reperfusion injury for example redistributes caveolin-3 and down-regulates expression of caveolin-1 within the heart [67]. Infusion of a peptide mimicking the caveolin scaffolding domain (CSD) of caveolin-1 into ischaemic/reperfused hearts results in the recovery of cardiac function, suggesting therapeutic potential [68]. Similarly, injection of a chimeric peptide consisting of the CSD of caveolin-1 fused to a cellular internalization sequence into mice selectively inhibits endothelial nitric oxide (eNOS) activity, decreasing inflammation-induced vascular leak [69]. CSD peptides thus have potent biological activity *in vivo* and small-molecule mimicry of the CSD has been suggested as having therapeutic potential for the treatment of pulmonary hypertension, atherosclerosis, inflammation, tissue fibrosis, and ischaemia/reperfusion injury [69-72]. Research interest in these small cellular compartments is thus likely to continue to expand in future years.

References

[1] Parton, R. G., and del Pozo, M. A. (2013) *Nat. Rev. Mol. Cell Biol.* 14, 98-112

[2] Palade, G. E. (1953) *Journal of Applied Physics* 24, 1424-1424

[3] Yamada, E. (1955) *Journal of Biophysical and Biochemical Cytology* 1, 445-&

[4] Parton, R. G., and Simons, K. (2007) *Nat. Rev. Mol. Cell Biol.* 8, 185-194

[5] Razani, B., Woodman, S. E., and Lisanti, M. P. (2002) *Pharmacol. Rev.* 54, 431-467

[6] Foster, L. J., de Hoog, C. L., and Mann, M. (2003) *Proc. Natl. Acad. Sci. U. S. A.* 100, 5813-5818

[7] Patel, H. H., Murray, F., and Insel, P. A. (2008) *Annu. Rev. Pharmacol. Toxicol.* 48, 359-391

[8] Andersen, O. S., and Koeppe, R. E. (2007) *Annual Review of Biophysics and Biomolecular Structure* 36, 107-130

[9] Pan, J. J., Tristram-Nagle, S., and Nagle, J. F. (2009) *Physical Review E* 80

[10] Rawicz, W., Smith, B. A., McIntosh, T. J., Simon, S. A., and Evans, E. (2008) *Biophysical Journal* 94, 4725-4736

[11] Okamoto, T., Schlegel, A., Scherer, P. E., and Lisanti, M. P. (1998) *J. Biol. Chem.* 273, 5419-5422

[12] Garg, V., Jiao, J. D., and Hu, K. L. (2009) *Cardiovascular Research* 82, 51-58

[13] Schwarzer, S., Nobles, M., and Tinker, A. (2010) *J. Biol. Chem.* 285, 27817-27826

[14] Cui, S., Ho, W. K., Kim, S. T., and Cho, H. (2010) *J. Biol. Chem.* 285, 41732-41739

[15] Alday, A., Urrutia, J., Gallego, M., and Casis, O. (2010) *Channels* 4, 168-178

[16] Cole, W. C., McPherson, C. D., and Sontag, D. (1991) *Circ.Res.* 69, 571-581

[17] Suzuki, M., Sasaki, N., Miki, T., Sakamoto, N., Ohmoto-Sekine, Y., Tamagawa, M., Seino, S., Marban, E., and Nakaya, H. (2002) *J. Clin. Invest.* 109, 509-516

[18] Cho, H., Lee, D., Lee, S. H., and Ho, W. K. (2005) *Proc. Natl. Acad. Sci. U. S. A.* 102, 4643-4648

[19] Yamamoto, M., Toya, Y., Schwencke, C., Lisanti, M. P., Myers, M. G., and Ishikawa, Y. (1998) *J. Biol. Chem.* 273, 26962-26968

[20] Rahman, A., and Sward, K. (2009) *Acta Physiologica* 195, 231-245

[21] Couet, J., Li, S. W., Okamoto, T., Ikezu, T., and Lisanti, M. P. (1997) *J. Biol. Chem.* 272, 6525-6533

[22] Byrne, D. P., Dart, C., and Rigden, D. J. (2012) *Plos One* 7

[23] Collins, B. M., Davis, M. J., Hancock, J. F., and Parton, R. G. (2012) *Developmental Cell* 23, 11-20

[24] Sampson, L. J., Hayabuchi, Y., Standen, N. B., and Dart, C. (2004) *Circ.Res.* 95, 1012-1018

[25] Sampson, L. J., Davies, L. M., Barrett-Jolley, R., Standen, N. B., and Dart, C. (2007) *Cardiovascular Research* 76, 61-70
[26] Jiao, J. D., Garg, V., Yang, B. F., Elton, T. S., and Hu, K. L. (2008) *Hypertension* 52, 499-506
[27] Quayle, J. M., Nelson, M. T., and Standen, N. B. (1997) *Physiol. Rev.* 77, 1165-1232
[28] Clapp, L. H., and Tinker, A. (1998) *Curr. Opin. Nephrol. Hypertens.* 7, 91-98
[29] Yokoshiki, H., Sunagawa, M., Seki, T., and Sperelakis, N. (1998) *Am. J. Physiol.-Cell Physiol.* 43, C25-C37
[30] Cole, W. C., and Clement-Chomienne, O. (2003) *Journal of Cardiovascular Electrophysiology* 14, 94-103
[31] Davies, L. M., Purves, G. I., Barrett-Jolley, R., and Dart, C. (2010) *J. Physiol.-London* 588, 3255-3266
[32] Aguilar-Bryan, L., Clement, J. P., Gonzalez, G., Kunjilwar, K., Babenko, A., and Bryan, J. (1998) *Physiol. Rev.* 78, 227-245
[33] Aguilar-Bryan, L., and Bryan, J. (1999) *Endocrine Reviews* 20, 101-135
[34] Seino, S. (1999) *Annual Review of Physiology* 61, 337-362
[35] Garg, V., Sun, W., and Hu, K. L. (2009) *Biochemical and Biophysical Research Communications* 385, 472-477
[36] Puddu, A., Salani, B., Cordera, R., Viviani, G. L., and Maggi, D. (2008) *Biochemical and Biophysical Research Communications* 375, 235-237
[37] Ashcroft, F. M., and Gribble, F. M. (1999) *Diabetologia* 42, 903-919
[38] Nevins, A. K., and Thurmond, D. C. (2006) *J. Biol. Chem.* 281, 18961-18972
[39] Ye, B., Balijepalli, R. C., Foell, J. D., Kroboth, S., Ye, Q., Luo, Y. H., and Shi, N. Q. (2008) *Biochemistry* 47, 12312-12318
[40] Lin, J. J., Lin, S. G., Choy, P. C., Shen, X. Z., Deng, C. Y., Kuang, S. J., Wu, J., and Xu, W. C. (2008) *Biochemistry and Cell Biology-Biochimie Et Biologie Cellulaire* 86, 405-415
[41] Richter, T., Floetenmeyer, M., Ferguson, C., Galea, J., Goh, J., Lindsay, M. R., Morgan, G. P., Marsh, B. J., and Parton, R. G. (2008) *Traffic* 9, 893-909
[42] Guo, J., Wang, T. Z., Li, X., Shallow, H., Yang, T. H., Li, W. T., Xu, J. M., Fridman, M. D., Yang, X. L., and Zhang, S. T. (2012) *J. Biol. Chem.* 287, 33132-33141
[43] McEwen, D. P., Li, Q., Jackson, S., Jenkins, P. M., and Martens, J. R. (2008) *Mol. Pharmacol.* 73, 678-685

[44] Alioua, A., Lu, R., Kumar, Y., Eghbali, M., Kundu, P., Toro, L., and Stefani, E. (2008) *J. Biol. Chem.* 283, 4808-4817

[45] Brainard, A. M., Miller, A. J., Martens, J. R., and England, S. K. (2005) *Am. J. Physiol.-Cell Physiol.* 289, C49-C57

[46] Lin, M. T., Adelman, J. P., and Maylie, J. (2012) *Am. J. Physiol.-Cell Physiol.* 303, C318-C327

[47] Vatta, M., Ackerman, M. J., Ye, B., Makielski, J. C., Ughanze, E. E., Taylor, E. W., Tester, D. J., Balijepalli, R. C., Foell, J. D., Li, Z. H., Kamp, T. J., and Towbin, J. A. (2006) *Circulation* 114, 2104-2112

[48] Vaidyanathan, R., Vega, A. L., Song, C. H., Zhou, Q., Tan, B. H., Berger, S., Makielski, J. C., and Eckhardt, L. L. (2013) *J. Biol. Chem.* 288, 17472-17480

[49] Seino, S., and Miki, T. (2003) *Prog. Biophys. Mol. Biol.* 81, 133-176

[50] Flagg, T. P., Enkvetchakul, D., Koster, J. C., and Nichols, C. G. (2010) *Physiol. Rev.* 90, 799-829

[51] Garlid, K. D., Costa, A. D. T., Quinlan, C. L., Pierre, S. V., and Dos Santos, P. (2009) *Journal of Molecular and Cellular Cardiology* 46, 858-866

[52] Di Lisa, F., and Bernardi, P. (2006) *Cardiovascular Research* 70, 191-199

[53] Quinlan, C. L., Costa, A. D. T., Costa, C. L., Pierre, S. V., Dos Santos, P., and Garlid, K. D. (2008) *Am. J. Physiol.-Heart Circul. Physiol.* 295, H953-H961

[54] Romanenko, V. G., Rothblat, G. H., and Levitan, I. (2002) *Biophysical Journal* 83, 3211-3222

[55] Romanenko, V. G., Fang, Y., Byfield, F., Travis, A. J., Vandenberg, C. A., Rothblat, G. H., and Levitan, I. (2004) *Biophysical Journal* 87, 3850-3861

[56] Li, H., and Papadopoulos, V. (1998) *Endocrinology* 139, 4991-4997

[57] Rosenhouse-Dantsker, A., Logothetis, D. E., and Levitan, I. (2011) *Biophysical Journal* 100, 381-389

[58] Singh, A. K., McMillan, J., Bukiya, A. N., Burton, B., Parrill, A. L., and Dopico, A. M. (2012) *J. Biol. Chem.* 287, 20509-20521

[59] Ramu, Y., Xu, Y. P., and Lu, Z. (2006) *Nature* 442, 696-699

[60] Bock, J., Szabo, I., Gamper, N., Adams, C., and Gulbins, E. (2003) *Biochemical and Biophysical Research Communications* 305, 890-897

[61] Ganapathi, S. B., Fox, T. E., Kester, M., and Elmslie, K. S. (2010) *Am. J. Physiol.-Cell Physiol.* 299, C74-C86

[62] Fu, Y., Moore, X. L., Lee, M. K. S., Fernandez-Rojo, M. A., Parat, M. O., Parton, R. G., Meikle, P. J., Sviridov, D., and Chin-Dusting, J. P. F. (2012) *Arterioscler. Thromb. Vasc. Biol.* 32, E117-+
[63] Joshi, B., Bastiani, M., Strugnell, S. S., Boscher, C., Parton, R. G., and Nabi, I. R. (2012) *J. Cell Biol.* 199, 425-435
[64] Pires, P. W., Ramos, C. M. D., Matin, N., and Dorrance, A. M. (2013) *Am. J. Physiol.-Heart Circul. Physiol.* 304, H1598-H1614
[65] Grayson, T. H., Ohms, S. J., Brackenbury, T. D., Meaney, K. R., Peng, K., Pittelkow, Y. E., Wilson, S. R., Sandow, S. L., and Hill, C. E. (2007) *Bmc Genomics* 8
[66] Patel, H. H., Zhang, S., Murray, F., Suda, R. Y. S., Head, B. P., Yokoyama, U., Swaney, J. S., Niesman, I. R., Schermuly, R. T., Pullamsetti, S. S., Thistlethwaite, P. A., Miyanohara, A., Farquhar, M. G., Yuan, J. X. J., and Insel, P. A. (2007) *Faseb Journal* 21, 2970-2979
[67] Ballard-Croft, C., Locklar, A. C., Kristo, G., and Lasley, R. D. (2006) *Am. J. Physiol.-Heart Circul. Physiol.* 291, H658-H667
[68] Young, L. H., Ikeda, Y., and Lefer, A. M. (2001) *Am. J. Physiol.-Heart Circul. Physiol.* 280, H2489-H2495
[69] Bucci, M., Gratton, J. P., Rudic, R. D., Acevedo, L., Roviezzo, F., Cirino, G., and Sessa, W. C. (2000) *Nat. Med.* 6, 1362-1367
[70] Jasmin, J. F., Mercier, I., Dupuis, J., Tanowitz, H. B., and Lisanti, M. P. (2006) *Circulation* 114, 912-920
[71] Rodriguez-Feo, J. A., Hellings, W. E., Moll, F. L., De Vries, J., van Middelaar, B. J., Algra, A., Sluijter, J., Velema, E., van der Broek, T., Sessa, W. C., De Kleijn, D. P. V., and Pasterkamp, G. (2008) *Plos One* 3
[72] Tourkina, E., Richard, M., Gooz, P., Bonner, M., Pannu, J., Harley, R., Bernatchez, P. N., Sessa, W. C., Silver, R. M., and Hoffman, S. (2008) *Am. J. Physiol.-Lung Cell. Mol. Physiol.* 294, L843-L861

In: Lipid Rafts
Editor: Dan Sillence
ISBN: 978-1-62948-961-2

Chapter IV

The Role of Caveolin-Enriched Microdomains in the Regulation of Cellular Signalling and Energy Homeostasis

Christopher Lipina, Mitchell H. Whyte, Clare Stretton and Harinder S. Hundal
Division of Cell Signalling & Immunology, College of Life Sciences, University of Dundee, Dundee, UK

Abstract

Lipid rafts are nanoscale regions of the plasma membrane with distinct structural compositions that function to modulate a diverse array of biological processes. In accordance with this, specialised lipid rafts termed caveolae have been implicated as key signalling platforms for various pathways and processes involved in the modulation of energy homeostasis. This chapter aims to highlight and review experimental evidence which supports a role for lipid rafts/caveolae in coordinating signalling events and processes that are crucial for the maintenance of normal glucose and lipid homeostasis. In particular, the involvement of these membrane microdomains in the regulation of insulin action, pro-inflammatory signalling, mitochondrial function and lipid metabolism will be discussed, as well as their potential role in the pathogenesis of human metabolic disorders.

Introduction

The plasma membrane provides a natural barrier which functions to tightly regulate the movement of biologically active molecules into or out of cells. This selective permeability is essential for regulating processes that are required to maintain proper cell function such as the modulation of cellular trafficking, intracellular signalling, and the preservation of ion gradients. At their most basic level, cell membranes are considered to consist of a fluid lipid bilayer predominantly made up of phospholipids and other amphipathic molecules. However, it is now widely recognised that the plasma membrane is not uniform in nature but rather consists of specialised domains each with their own distinct structural profile. One such group of membrane domains are lipid rafts. These nanometre-scale membrane regions are often enriched in sterols, saturated phospholipids and sphingolipids, and due to this tend to be more ordered and less fluid than the surrounding lipid bilayer. Moreover, specific proteins can localise within these domains that are involved in mediating their formation and/or biological functions [1].

One important consideration in the study of lipid rafts and their biology is the dynamic nature of their assembly/disassembly – which can sometimes be measured on a nanosecond timescale. Because of this, studies investigating the function of lipid rafts commonly implement one or more biochemical strategies alongside high resolution microscopy. For example, imaging by immunogold labelling and transmission electron microscopy is now commonly used to detect raft like regions within the plasma membrane through targeting specific lipid or protein components [2].

To date, two main types of lipid raft have been proposed to exist: planar rafts and caveolae. Planar rafts lie continuous within the plane of the membrane and despite their distinct lipid/protein composition, they lack any distinguishable morphological features. In contrast, caveolae form characteristic flask-like or omega-shaped invaginations of the plasma membrane that provide a favourable microenvironment for the modulation of specific signalling pathways and processes. The following section describes in more detail what makes caveolae such a specialised lipid raft domain.

"Caveolae Intracellularis"

The term "*Caveolae intracellulares*" was first used to describe very small pit-like depressions observed in cell membranes by electron microscopic visualization of mouse gall bladder epithelia [3]. Indeed, since then, the term caveolae has become synonymous with flask or omega-shaped invaginations (50 -100 nm diameter) of the plasma membrane (see Figure 1). The abundance of these subcellular structures can vary markedly in different cell types from high caveolar content in adipocytes and endothelial cells (which can constitute up to 30% of plasma membrane surface) in contrast to their sparse distribution (less than 5% of plasma surface membrane surface) within fibroblasts [4].

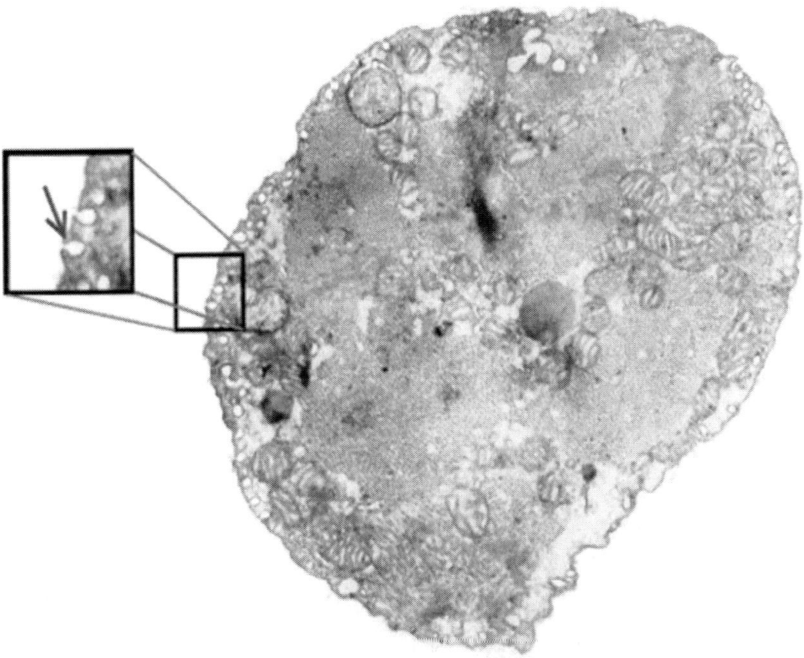

Figure 1. Electron micrograph of an adult rat ventricular myocyte showing the presence of caveolae (arrow in inset) at the cell periphery. Figure adapted from Besse IM & Mitchell CC *et al* 2011 [151].

In terms of their structural composition, caveolae are membrane regions that are enriched in cholesterol and sphingolipids, as well as a number of key proteins that are crucial for their formation and function. One such family of proteins is the caveolins which act as coat proteins for caveolae and are essential for their formation. Importantly, depletion of caveolin isoforms either through gene silencing or gene knockout methods has been shown to cause a profound loss of morphologically detectable caveolar-like structures in a number of different cell types [5-10]. Three caveolin subtypes have been identified in mammalian cells – caveolin-1, caveolin-2 and caveolin-3 –which are transcribed from 3 distinct genes (CAV1, CAV2 and CAV3, respectively). Furthermore, two variants of caveolin-1 termed caveolin-1α and caveolin-1β can arise due to alternative initiation or splicing of the Cav1 gene. As a result, the caveolin-1β isoform lacks the first 31 amino acids present in caveolin-1α [11-14]. All caveolin isoforms contain membrane spanning domains that allow these proteins to adopt their characteristic hairpin loop conformation leaving both the N and C-terminal regions exposed to the cytoplasm [15].

Caveolin-1 is found to be widely expressed and is the predominant isoform in a number of cell types including adipocytes and endothelial cells [9, 16]. Although the function of caveolin-2 remains poorly understood, it is known to co-express and interact with caveolin-1 to form large molecular mass hetero-oligomers that are necessary for the formation of deep plasma membrane caveolae [17]. Indeed, caveolin-2 may function as some form of accessory protein for caveolin-1. Caveolin-3 is found to be predominantly expressed in myocytes [18]. Unlike caveolin-2, caveolin-3 can form homo-oligomeric complexes and does not require caveolin-1 to drive caveolar formation [10, 19].

In addition to caveolins, cholesterol is another key component of caveolae that is required for their formation [20, 21]. Indeed, due to the ability of caveolin isoforms to bind cholesterol with high affinity [4], this key property is likely to facilitate their localisation within cholesterol-enriched membrane regions such as caveolae. Furthermore, cholesterol has also been reported to upregulate caveolin gene transcription [21]. Accordingly, cellular depletion of cholesterol using chemical agents such as methyl-β-cyclodextrin (MβCD) has been shown to promote the dissolution of caveolin-enriched microdomains (CEMs) (see Figures 2 and 3) [22-24]. In the case of MβCD, this particular compound acts to sequester cholesterol from cell membranes via interactions of the lipid with the hydrophobic core of its cup-like structure (see Figure 3). Other compounds which produce similar cholesterol sequestering effects include filipin and nystatin [25-27] .

All caveolin isoforms possess a caveolin scaffolding domain (CSD) which mediates their interaction with proteins containing a specific recognition motif (ϕXϕXXXXϕ and ϕXXXXϕXXϕ where ϕ is the aromatic residue Phe, Tyr or

Trp and X is any residue). Importantly, a number of proteins have not only been identified which contain such CSD-binding motifs (also termed caveolin binding motifs), but have also been shown to directly interact with caveolins including various receptors, ion channels, structural proteins and enzymes (see Table 1). Intriguingly, it has been estimated that up to 30% of the human proteins may contain these putative caveolin binding motifs [28]. In accordance with this, caveolae have been shown to coordinate and/or influence various cellular processes including nutrient transport, membrane trafficking as well as a multitude of signal transduction pathways [29, 31].

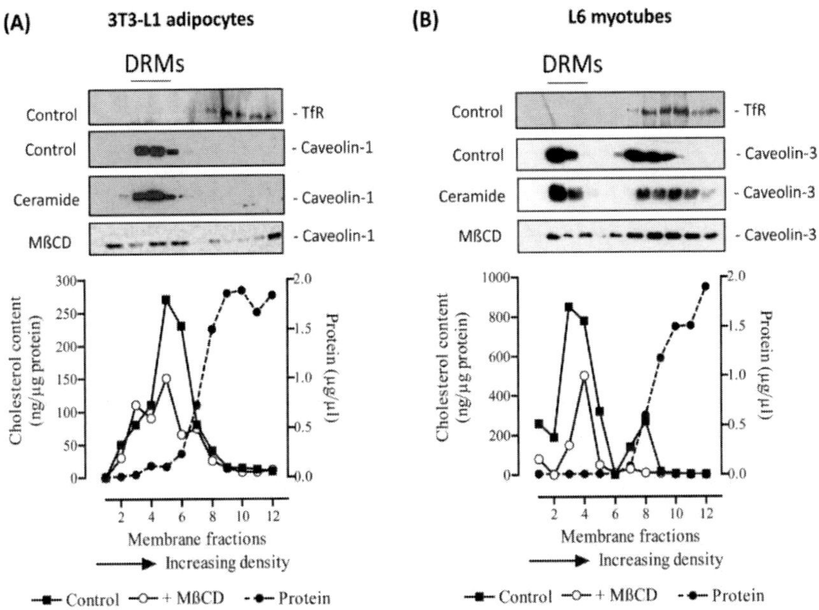

Figure 2. Protein and lipid composition of detergent resistant membrane fractions (DRMs) isolated from 3T3-L1 adipocytes and rat skeletal L6 myotubes treated with or without the cholesterol depleting agent methyl-β-cyclodextrin (MβCD) as indicated. The increased abundance of caveolin-1 (in adipocytes) or caveolin-3 (in L6 myotubes) as well as higher cholesterol content is evident within isolated DRM fractions. Cholesterol depletion using MβCD dramatically alters the composition of DRMs in both cell types. Figure adapted from Hajduch *et al,* 2008 [23].

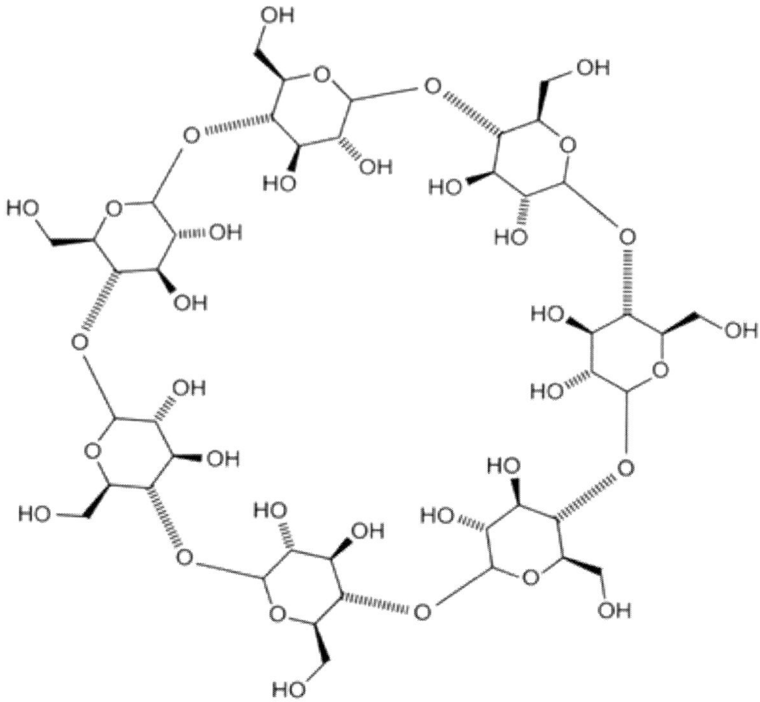

Figure 3. Structure of β-cyclodextrin illustrating its cup-like hydrophobic core which acts to bind and sequester cholesterol (Adapted from De Oliveira *et al*, 2011[152]).

Table 1. Examples of proteins with different biological functions that have been reported to interact with caveolin isoforms. Table includes the location of the caveolin binding motif (CBM) (based on human sequence), protein function and relevant citations

Protein Name	**Location of CBM within protein sequence**	**Protein Function**	**References**
Protein kinase Cζ	428-435	Membrane/Cytosolic Signalling	[154]
Insulin Receptor	1220-1227	Hormonal/Growth factor receptor	[62, 155]
Aquaporin-1	210-217	Water channel	[156]
Protein Phosphatase 5	323-335	Protein Dephosphorylation	[157]
Fatty Acid Synthase	1506-1514	Lipid Metabolism	[158]
Toll-like Receptor 4 (TLR4)	1146-1153	Inflammatory Signalling/Response	[109]

Caveolin proteins also undergo a number of different post-translational modifications that can impact on their functional interactions. For example, palmitoylation of both caveolin-1 and caveolin-3 on specific cysteine residues located within their C-terminal regions has been found to enhance their binding affinity for cholesterol, as well as promoting the stabilization of caveolin-1 oligomers [32, 33]. In addition, the phosphorylation of caveolin-1 can also modulate its interaction with other molecules [34]. For example, phosphorylation of caveolin-1 at Tyr 14 by members of the Src family of protein kinases, such as Src and Fyn, has been reported to alter its affinity for interacting proteins as well as influencing caveolin-mediated cellular processes [35, 36].

In addition to caveolins, another distinct family consisting of four related proteins that have also been implicated in caveolar organization and function are the cavins. Like caveolins, cavins also play a key role in the formation of caveolae by serving as adaptor molecules for caveolin coat proteins. Liu and colleagues first demonstrated that mice lacking cavin-1 (also known as PTRF; polymerase I and transcript release factor) were devoid of morphologically detectable caveolae in all tissues examined [37]. Furthermore, a study by Hill and colleagues showed that forced expression of cavin-1 in prostate cancer PC3 cells was sufficient to induce formation of caveolae [38]. Cavin-1 has also been reported to direct the formation of flask-shaped caveolae through its interactions with cavin-2 [37, 39, 40].

Whilst cavin-3 has not been shown to participate in caveolar formation, this particular isoform has been reported to co-immunoprecipitate with caveolin-1 and can modulate caveolar-dependent vesicle trafficking [41]. Although little is known about the function of cavin-4, it is found to be predominantly expressed in skeletal and cardiac muscle where it has been shown to positively regulate muscle biogenesis [42, 43]. Cavins have also been proposed to regulate a number of caveolae-dependent signalling pathways and processes [37, 44, 45].

In addition to caveolins and cavins, flotillin proteins are also found to localise within CEMs. Two flotillin members have been identified, flotillin-1 and flotillin-2 (encoded by the FLOT1 and FLOT2 genes respectively), which co-assemble to form hetero-oligomeric structures with each other and with caveolins. Whether flotillins are directly involved in caveolar formation is not yet understood, however, work performed in insect cells has demonstrated that forced expression of flotillin-1 is sufficient to promote the formation of caveolae-like vesicles [46]. Importantly, similar to caveolins, flotillins are also known to play a key role in regulating a number of caveolar-dependent processes including endocytosis, nutrient transport and growth factor mediated signaling [47-49].

Isolation of Lipid Raft/Caveolar Microdomains

The most common method used to isolate components of lipid rafts and caveolin-enriched microdomains (CEMs) involves treating cell extracts with non-ionic detergents such as Triton X-100 at cold temperatures which serves to solubilize the bulk of the plasma membrane bilayer whilst leaving many detergent-insoluble components of lipids rafts/CEMs intact. This is largely due to the high abundance of sphingolipids and cholesterol that constitute these microdomains, as well as the detergent-resistant nature of glycophosphatidylinositol (GPI)-anchored proteins that reside within these membrane regions. Once cell membranes are homogenized in cold detergent, detergent resistant membrane (DRM) fractions are then isolated by sucrose density gradient centrifugation. During high speed centrifugation, homogenized membrane components are forced to move towards higher density fractions whilst the less dense DRM containing fractions translocate to a lower density sucrose interface (for example at the 5-30% sucrose interface). Importantly, analysis of these resulting DRM fractions consistently reveals the presence of a number of components associated with lipid rafts/CEMs and because of this, DRMs are often considered to represent their biochemical equivalent.

It should be noted, however, that caution must be exercised when interpreting data derived from isolated DRM fractions. For example, isolation techniques based on detergent insolubility can in some instances be considered to be artifactual in nature and may not always accurately represent their true composition within the plasma membrane – in particular when detergent isolation is performed at lower temperatures (~4°C). Indeed, applying alternative detergents (for example, comparing Triton X-100 with Brij detergents and CHAPSO) has been shown to produce altered distribution profiles of raft proteins, cholesterol and various lipid species within corresponding fractions following gradient centrifugation [50, 51]. This is likely due, in part, to the varying extents to which these different detergents act to solubilise membrane proteins and concentrate sphingolipids and cholesterol.

This chapter aims to highlight and review experimental evidence pertaining to the involvement of caveolae/lipid rafts in processes that are important for the maintenance of energy homeostasis. In particular, the potential role of these membrane microdomains in regulating insulin-induced signalling, pro-inflammatory responses, adipocyte function and mitochondrial homeostasis will be discussed. In addition, any implications of experimental findings made with regards to the pathogenesis of metabolic disease in humans will also be highlighted.

Insulin-Regulated Signal Transduction

Coordinated regulation of the uptake and utilisation of glucose, the preferential fuel for nearly all cell types, is crucial for the maintenance of normal cellular function and survival. One key hormone that plays a major role in the modulation of circulating blood glucose, primarily through promoting glucose clearance, is insulin. Indeed, any impairment in the secretion of insulin from pancreatic beta cells and/or reduced sensitivity of its target tissues, termed insulin resistance, can lead to the development of hyperglycaemia and its associated pathologies such as, for example, cardiovascular disease and neuropathy. Importantly, such adverse health effects are often linked to obesity and Type II diabetes which are associated with insulin resistance and hyperglycaemia.

The binding of insulin to its receptor stimulates insulin receptor kinase activity which promotes tyrosine phosphorylation of various substrates including the IRS (insulin receptor substrate) family of proteins (IRS-1, IRS-2 and IRS-4). As a result, phosphotyrosine motifs present within these IRS isoforms mediate the recruitment and activation of PI3K (phosphoinositide 3-kinase) which catalyzes the generation of 3-phosphoinositides such as phosphatidylinositol (3,4,5) trisphosphate (PIP3) within the plasma membrane. Consequently, this increase in PIP3 facilitates the activation of several key components of the classical insulin signalling pathway including protein kinase B (PKB)(also known as Akt) [52]. In particular, the interaction of the N-terminal PH (pleckstrin homology) domain of PKB/Akt with PIP3 induces conformational changes in this kinase which permits the phosphorylation of its two regulatory sites, Thr308 and Ser473, by PDK1 (3-phosphoinositide dependent kinase) and mTORC2, respectively [53-55].

Once activated, PKB/Akt can target a number of substrates which are involved in mediating the physiological effects of insulin. For example, this includes the cellular uptake of glucose which is facilitated by glucose transporters such as GLUT4. More specifically, PKB/Akt has been proposed to stimulate glucose transport through its ability to suppress the Rab GTPase activating protein (GAP) function of AS160 thereby leading to increased GLUT4 vesicle trafficking [56]. Furthermore, PKB/Akt also mediates the inhibition of GSK3 (glycogen synthase kinase 3) which contributes to insulin-induced activation of glycogen synthesis [57].

The following sections discuss the role of CEMs in insulin signal transduction, as well as their potential involvement in the development of insulin resistance.

The Role of Cems in Insulin-Induced Signalling

Activation of the PI3K-PKB/Akt signalling axis is crucial for conveying numerous insulin induced metabolic responses. Importantly, several lines of evidence implicate caveolar microdomains as key modulators of insulin-directed signalling. Firstly, a number of proteins that participate in insulin signal transduction (for example the insulin receptor, IRS-1, PI3K, PDK1 and PKB/Akt) have either been detected within CEMs and/or shown to directly interact with caveolin isoforms [58-61] (see Table 2). Also, consistent with experimental evidence demonstrating their interaction with caveolin isoforms, sequence analysis also reveals that several of these proteins possess one or more putative caveolin-binding domains (see Table 2). For example, the insulin receptor which contains three such caveolin-binding motifs has been reported to localise at the neck regions of caveolae and can directly interact with caveolin-1 [61-64].

Secondly, a number of studies have implemented either gene silencing or knockout strategies to disrupt caveolar function [65, 66]. Accordingly, silencing the expression of caveolin-1 has been shown to markedly reduce insulin-dependent glucose uptake within skeletal muscle cells and/or adipocytes [16, 63]. Furthermore, immortalised mouse skeletal myotubes deficient for caveolin-3 display suppressed insulin-induced PI3K and PKB/Akt activation, concomitant with impaired GLUT4 translocation [47]. Several *in vivo* studies also demonstrate an important role for caveolae in the modulation of insulin sensitivity. For example, mice deficient for either caveolin-1 or caveolin-3 are found to exhibit glucose intolerance and insulin resistance [67,

68]. Importantly, the reduced insulin sensitivity exhibited by caveolin-3 knockout mice has been linked to attenuated activation of the insulin receptor and downstream targets including PKB/Akt [68, 69]. Similar to caveolin-1/3 knockout animals, mice lacking cavin-1 are also found to be glucose intolerant and hyperinsulinemic, despite being of normal weight and having reduced fat mass [37].

Table 2. List of caveolin-interacting proteins implicated in the modulation of insulin signal transduction. Included in the table are the location of putative caveolin binding motifs (CBM) (numbers refer to human sequences) identified in the proteins and references to studies which have demonstrated either their interaction with caveolin isoforms and/or their localisation in caveolin-enriched membrane microdomains. X refers to any amino acid residue

Protein	Caveolin Binding Motif	Position of sequence(s)	References
Insulin Receptor	FXYXXXXF FXYXXXXF WXFXXXXW	116-123 502-509 1220-1227	[62, 155]
IRS1	No Match	-	[159]
PI3-K – regulatory Subunit	No Match	-	[159]
PI3-K – Catalytic Subunit	FXYXXXXF FXXXXYXXF	214-221 798-806	[159]
PDK1	YXFXXXXF FXXXXYXXF	141-149 299-306	[60, 160]
PKB/Akt	No Match	-	[60]
PKCξ	YXFXXXXW	428-435	[23]
PTEN	FXFXXXXF	271-278	[23]
PTP-1B	FXXXXWXXF	174-182	[87, 161]

Consistent with these findings, targeted gene transfer of caveolins has been reported to improve insulin sensitivity [68-70]. For example, the insulin resistance displayed by caveolin-3 deficient mice can be reversed by adenovirus-mediated genetic transfer of caveolin-3 into the solei of these animals [69]. Similarly, the up-regulation of caveolin-3 in liver by adenovirus mediated gene transfer has likewise been shown to enhance hepatic insulin sensitivity in KKA(y) diabetic mice [70]. Furthermore, a study by Trajkovski and colleagues demonstrated that microRNA silencing of miR103 and miR107 (whose levels are both upregulated in obese mice) leads to increased expression of caveolin-1 in adipocytes concomitant with stabilisation of the insulin receptor and improved insulin sensitivity [71].

There are a number of reasons that may explain why disrupting caveolar integrity/function impacts negatively upon insulin action. Firstly, because various components of insulin signalling have been shown to be localised within caveolae including the insulin receptor itself, it may be the case that these membrane microdomains act as key scaffolding points for coordinating the assembly of signalling complexes that are crucial for insulin signal transduction. In addition, caveolar microdomains may also create an environment which allows for the stabilisation of these components. To support this, genetic depletion of caveolin-1, caveolin-3 or cavin-1 has been shown to promote insulin receptor protein destabilisation in adipocytes and/or skeletal muscle [16, 37, 67, 68]. Similarly, genetic deletion of caveolin-1 in mouse embryonic fibroblasts (MEFs) has been reported to downregulate IRS-1 protein abundance, which can be restored by reintroduction of caveolin-1 [72]. Together, these observations suggest that caveolae may serve to protect against degradation of these key insulin signalling components although the mechanism(s) by which they do so remains unclear. It must also be noted that cholesterol depletion of adipocytes by application of methyl-β-cyclodextrin has similarly been found to attenuate PKB/Akt activation in response to insulin, although, unlike caveolin gene silencing, can do so without altering insulin receptor protein expression and/or its activation [59, 73, 74].

Another possibility relates to the internalization and recycling of the insulin receptor following its activation. Autophosphorylation of the insulin receptor in response to insulin binding is known to initiate endocytosis of the ligand-receptor complex and its subsequent transport to endosomes. Unbound receptors are then transported back to the plasma membrane in vesicles whilst ligands (i.e. insulin) undergo lysosomal degradation. Importantly, Fagerholm and colleagues first demonstrated in primary adipocytes that the autophosphorylated insulin receptor becomes rapidly endocytosed *via* a process that is caveolae-dependent [75]. The authors of this study also reported the colocalisation of the insulin receptor and caveolin-1 within endosomal vesicles. Therefore, it is possible that impairing caveolar function may interfere with insulin receptor recycling and therefore perturb further signaling in response to insulin.

As well as modulating insulin signal transduction, caveolae may also play a key role in coordinating insulin-dependent glucose uptake. For example, studies performed in adipocytes have reported the localisation of the glucose transporter GLUT4 within CEMs which can be further increased in response to insulin stimulation [76-78]. Crucially, work by Fecchi and colleagues first demonstrated a profound impairment of insulin-induced GLUT4 translocation and glucose uptake when using immortalized skeletal muscle cells isolated from caveolin-3 null mice [47]. Such reductions in glucose transport capacity may, in part, be linked to impaired PI3K-PKB/Akt dependent signalling. However, expression of a dominant interfering mutant form of caveolin-3 has been reported to suppress insulin-induced GLUT4 translocation without altering PI3K activity in adipocytes [79]. In this particular case, failure of the small GTP binding protein TC10 to localise within the caveolin-enriched membrane domains was proposed to account for the loss in GLUT4 translocation. This is in agreement with previous work showing that insulin can promote GLUT4 translocation in adipocytes *via* activation of the Cbl-associated protein (CAP)/TC10 pathway which acts independently of PI3K [79, 80]. Therefore, caveolae may be involved in coordinating multiple pathways that can mediate GLUT4 vesicle trafficking and glucose transport.

It must be noted that noncaveolar microdomains have also been reported to modulate insulin receptor mediated signalling. For example, caveolin-2 has been shown to interact directly with the insulin receptor and can mediate its activation in caveolar-deficient 3T3-L1 adipocytes lacking caveolin-1 [81, 82]. Moreover, the insulin receptor has also been reported to localise within noncaveolar lipid raft domains enriched with homo-oligomeric cavin-2, where its insulin-induced activation remains unaffected by cholesterol depletion [81, 82]. Therefore, such noncaveolar microdomains may play some additional role in mediating insulin signal transduction, particularly under those conditions where caveolar integrity/function becomes compromised.

The Role of Caveolae in the Development of Insulin Resistance

Insulin desensitisation can arise in response to a variety of different obesity-related stimuli including sustained exposure to saturated fatty acids or inflammatory cytokines. Intriguingly, as well as positively modulating insulin receptor induced signalling, there is now growing evidence to suggest that caveolae may also participate in the development of insulin resistance. For example, CEMs have been implicated in promoting the insulin desensitising effects of the lipid intermediate ceramide which acts as a potent repressor of insulin-stimulated PKB/Akt activation [23, 83]. Intracellular ceramide can be generated through its *de novo* synthesis from the long chain saturated fatty acid palmitate (C16:0) and has been shown to partly mediate its ability to cause insulin resistance [84].

Previous work published by our group has demonstrated that ceramide-induced activation of atypical PKC isoforms (λ/ζ) is a major mechanism by which this lipid intermediate acts to suppress PKB/Akt activity [23, 84]. More specifically, ceramide promotes the association of PKC(λ/ζ) with PKB/Akt resulting in the phosphorylation and subsequent repression of PKB/Akt by PKC(λ/ζ) [23]. Importantly, this ceramide-induced inhibition of PKB/Akt coincides with the localisation of the PKC(λ/ζ)-PKB/Akt complex within CEMs [23]. Crucially, caveolar disruption through cholesterol depletion can prevent targeting of PKC (λ/ζ) to caveolar microdomains and subsequently ameliorate ceramide-induced inhibition of PKB/Akt. Consistent with this, adipocytes isolated from caveolin-1 deficient mice are found to exhibit greater resistance to the insulin desensitising effects of ceramide [23].

Besides acting to stimulate atypical PKC isoforms, ceramide has also been reported to inhibit PKB/Akt through its ability to activate protein phosphatase 2A (PP2A) – a serine/threonine phosphatase which acts to dephosphorylate PKB/Akt at both its key (Thr308 and Ser473) regulatory sites. Intriguingly, a study by Blouin and colleagues observed that in cells with low caveolar abundance such as 3T3-L1 pre-adipocytes, ceramide was more likely to cause PKB/Akt inhibition through a PP2A-dependent mechanism [85]. This is in contrast to cell types displaying relatively high caveolar abundance such as differentiated 3T3-L1 adipocytes where ceramide was more likely to repress PKB/Akt through PKC (λ/ζ) activation. Therefore, cellular caveolar abundance may be a key determinant of the mechanism by which ceramide acts to inhibit PKB/Akt. Furthermore, as well as repressing PP2A, ceramide has also been shown to promote the recruitment of the lipid 3'-phosphoinositide phosphatase PTEN (Phosphatase and Tensin homolog deleted on chromosome Ten) to caveolar microdomains, potentially through its direct association with caveolins [23, 86]. PTEN functions to reduce levels of PIP3 at the plasma membrane thereby attenuating PI3K-dependent signal transduction triggered in response to insulin. However, whether the localisation of PTEN within CEMs is required for its lipid phosphatase activity remains to be explored.

An alternative phosphatase implicated in modulating insulin sensitivity that has also been reported to associate with caveolin isoforms is PTP-1B (Protein Tyrosine Phosphatase 1B) [87]. PTP-1B impairs insulin action by dephosphorylating the insulin receptor at specific tyrosine residues leading to its reduced activation [88]. In accordance with this, PTP-1B inhibition has been shown to partially antagonise palmitate induced insulin resistance in C2C12 myotubes [89]. Interestingly, adenovirus-mediated gene transfer of caveolin-3 into the livers of obese diabetic mice has been reported to enhance hepatic insulin sensitivity concomitant with the inhibition of PTP-1B [70]. Therefore, increased PTP-1B localisation to CEMs may be a contributing factor towards obesity induced insulin resistance.

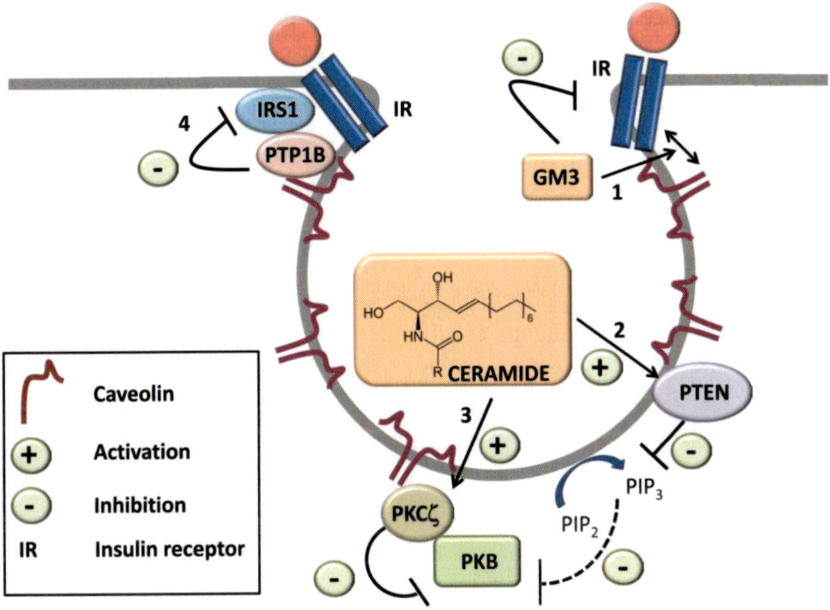

Figure 4. Schematic diagram illustrating the potential mechanisms by which caveolae may be involved in the pathogenesis of insulin resistance. (1) Accumulation of the ganglioside GM3 has been proposed to suppress insulin action through inducing the dissociation of the insulin receptor from caveolin-1. (2) The lipotoxic intermediate ceramide has been reported to stimulate activation of the lipid phosphatase PTEN and its recruitment to caveolar microdomains where it can act to impair insulin receptor induced signalling through suppressing PIP3 generation. (3) Alternatively, ceramide can impair insulin action through promoting the localisation of atypical PKC isoforms (λ/ζ) to caveolar microdomains thereby facilitating the association of PKC(λ/ζ) with PKB/Akt. This resulting interaction enables PKC(λ/ζ) to phosphorylate and inhibit PKB/Akt. (4) The protein phosphatase PTP-1B has been reported to localise within caveolar microdomains which may facilitate its ability to dephosphorylate and inactivate the insulin receptor and/or IRS-1.

As well as acting as a potent inducer of insulin resistance itself, ceramide can be further modified into other lipid intermediates that may also impact upon metabolic function. One such group of lipids that can be generated from ceramide are the gangliosides. In structural terms, gangliosides are ceramide-derived glycosphingolipids containing various carbohydrate modifications. One ganglioside in particular that has been implicated in the modulation of insulin sensitivity is GM3. A study by Yamashita and colleagues first reported that mice which lack the capacity to generate GM3 exhibit enhanced insulin sensitivity and are protected from high fat diet induced insulin resistance [90]. Interestingly, work by Kabayama and co-workers proposed that GM3 accumulation within caveolar microdomains can promote dissociation of the insulin receptor and caveolin-1, potentially driven by the competitive interaction of GM3 with the insulin receptor [64]. Therefore, obesity-induced increases in GM3 content may at least partly contribute to the development of insulin resistance through altering caveolin-dependent interactions [64, 92, 92].

Collectively, these findings suggest that caveolae and their components, which can positively modulate insulin action, may also act as crucial focal points for coordinating processes which lead to insulin desensitisation (see Figure 4). The following section describes how caveolae may further modulate insulin sensitivity and associated metabolic processes through regulating pro-inflammatory responses.

Role of Lipid Rafts/Caveolae in Obesity-Associated Inflammatory Signalling

It is now widely recognised that chronic inflammation is a key factor in the development of a number of pathological states including obesity-associated insulin resistance and metabolic dysfunction [93]. One key transcription factor involved in mediating cellular inflammatory responses is nuclear factor-κB (NF-κB). Under basal (i.e. non-stimulated) conditions, inactive NF-κB is sequestered within the cytoplasm by a family of inhibitory binding partners termed the IκB (inhibitor of κB) proteins. This association acts to prevent translocation of NF-κB into the nucleus and therefore suppress its transcriptional activity [94]. Various extracellular stimuli have been shown to promote activation of the serine/threonine kinase IκB-Kinase (IKK) which targets IκB [95]. Following phosphorylation by IKK, IκB isoforms become earmarked for proteasomal degradation thereby resulting in the release of active NF-κB into the nucleus where it can drive transcription of its target pro-inflammatory genes such as TNFα.

A chronic state of inflammation can be induced in response to several obesity-related factors including increased circulating levels of saturated fatty acids such as palmitate (C16:0) [96-99]. For example, sustained exposure of skeletal muscle cells to palmitate has been shown to stimulate the NF-κB dependent production of pro-inflammatory cytokines including TNF-α (tumour necrosis factor α). Importantly, these cytokines can promote insulin resistance through activating various stress-induced kinases such as c-jun terminal kinase (JNK) (see Figure 5) [100, 101]. Specifically, these stress-activated kinases have been proposed to impair insulin action at least partly through their ability to serine phosphorylate and inactivate IRS-1 [101, 102]. In accordance with this, mice which are genetically deficient in IKKβ, JNK1 or TNF-α all exhibit improved insulin sensitivity under conditions of diet-induced obesity [101, 103, 104].

Several studies propose that caveolae may regulate inflammatory responses either in a positive or negative manner. For example, Garrean and colleagues first demonstrated the attenuation of lipopolysaccharide (LPS) induced lung inflammation in mice lacking caveolin-1 [105]. In contrast, an enhanced inflammatory response has been reported in several different tissues of caveolin-1 deficient mice infected with *Pseudomonas aeruginosa* – concomitant with increased activation of NF-κB [106]. Consistent with this elevated inflammatory response, knockdown of caveolin-1 using a dominant negative strategy in lung epithelial cells was similarly found to promote increased pro-inflammatory cytokine production [107].

One feature of obesity induced inflammation that may be dependent upon functional caveolae is the ability of fatty acids to promote pro-inflammatory signalling through binding and activating specific members of the Toll-like receptor (TLR) family, in particular TLR2 and TLR4. Interestingly, a study by Chen and co-workers performed in endothelial cells demonstrated that palmitate can stimulate the recruitment of TLR4 into caveolin-1 containing membrane microdomains [108]. The same study also demonstrated that incubation with apolipoprotein A-I, a major protein component of high density lipoprotein, was able to block palmitate-induced NF-κB activation, concomitant with the attenuation of palmitate-induced TLR4 trafficking into lipid rafts [108]. In contrast, there is also evidence to support caveolae-mediated repression of TLR4 function. For example, the association between caveolin-1 and TLR4 has been shown to reduce production of TNFα and IL-6 in murine macrophages following treatment with LPS (lipopolysaccharide) – a known TLR4 agonist [109]. Furthermore, silencing the CAV1 gene in human mammary epithelial cells has been reported to enhance LPS-induced inflammatory signalling and gene expression [110].

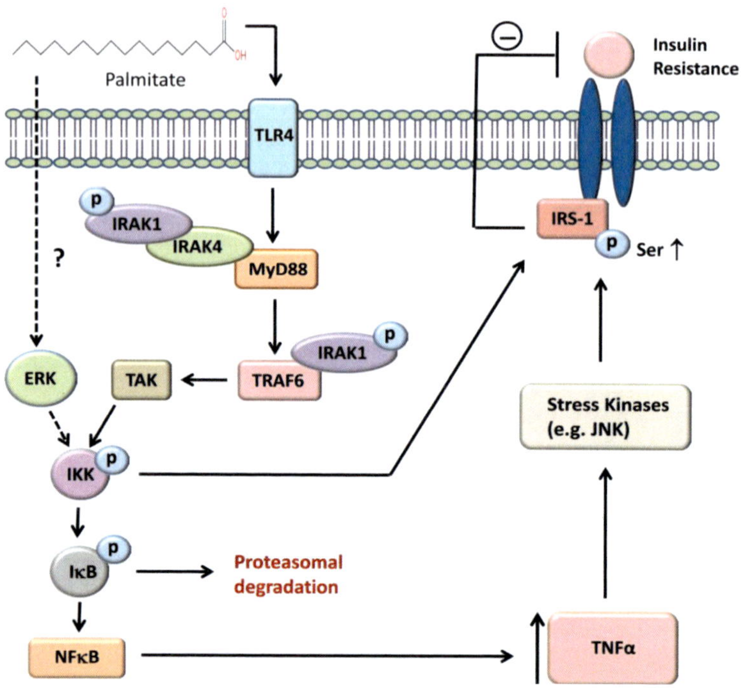

Figure 5. Schematic diagram illustrating how fatty acids such as palmitate may induce insulin desensitisation through their ability to stimulate pro-inflammatory signalling. Long chain saturated fatty acids such as palmitate (C16:0) can initiate NF-κB dependent pro-inflammatory signalling either by activating the mitogen activated protein kinase ERK, or alternatively by stimulating TLR4 dependent signaling [87, 153]. In either case, the subsequent activation of IKK (IκB kinase) permits it to target its substrate IκB, the inhibitory binding partner of NF-κB, thereby marking it for proteasomal degradation. Following degradation of IκBα, active unbound NF-κB becomes free to translocate into the nucleus where it can upregulate the transcription of pro-inflammatory target genes including TNFα. The subsequent production and secretion of pro-inflammatory cytokines permits them to act via their receptors on target tissues to stimulate various stress responsive kinases including JNK. These stress activated kinases then proceed to serine phosphorylate and subsequently impair the function of upstream insulin signalling components such as the IRS proteins. In addition, active IKK can itself act to suppress insulin signal transduction through its ability to serine phosphorylate IRS-1.

However, despite these findings, it remains unclear as to how caveolar function impacts upon pro-inflammatory signalling within peripheral tissues that play an important role in energy homeostasis (e.g. skeletal muscle and adipose tissue) – particularly in the obese state. Indeed, future work may involve comparing obesity-associated inflammation (i.e. cytokine expression and NF-κB signalling) in such tissues isolated from wild type versus caveolin-1 or -3 deficient mice.

Notably, however, caveolae have been implicated in mediating the insulin desensitising effects of pro-inflammatory cytokines. For example, it has been demonstrated that stimulating adipocytes with TNFα promotes dissociation of the insulin receptor from caveolae – a response which may contribute to the insulin desensitising effects of this cytokine [64, 92]. Importantly, blocking the intracellular synthesis of GM3 was found to ameliorate the insulin desensitising effects of TNFα within adipocytes [92]. Therefore, it may be the case that obesity-induced alterations in certain lipid and/or protein components of membrane microdomains (e.g. caveolins or sphingolipids) may not only impact directly upon NF-κB dependent pro-inflammatory signalling, but also instigate and/or augment metabolic perturbations that are associated with exposure to pro-inflammatory mediators.

Role of Caveolae in Adipocyte Function and Lipid Metabolism

In addition to regulating insulin signal transduction, caveolae and/or their components may also play an important role in coordinating other aspects of adipocyte function. Importantly, a study by Razani and colleagues first reported that mice deficient for caveolin-1 (CAV1), the predominant caveolin isoform in adipose tissue, were leaner and exhibited greater resistance to diet-induced obesity compared to wild type animals. This inability to gain fat mass occurred despite these animals being hyperphagic and having a normal capacity to absorb dietary fat [9]. More detailed analysis of adipose tissue isolated from caveolin-1 null mice showed not only a marked reduction in adipocyte number, but also identified histological abnormalities within individual adipocytes including significantly reduced lipid droplet formation [9].

In accordance with these *in vivo* observations, a number of subsequent studies have revealed that caveolins play an important role in coordinating triglyceride synthesis. Work by Ost and colleagues performed in primary adipocytes demonstrated that exogenous oleic acid (C18:1) can be taken up and converted into triglycerides within a specific subclass of caveolae [111]. Importantly, these caveolae were found to contain perilipin proteins which act to regulate lipid storage and packaging within adipocytes [111]. Intriguingly, although caveolins are predominantly located at the cell surface, they can redistribute to lipid droplets during adipocyte differentiation, or in response to fatty acid or cholesterol provision [112-114]. Indeed, the high binding affinity of caveolin-1 towards cholesterol most likely underlies the key role that this protein plays in maintaining sufficient cholesterol levels which are required for the formation of well-developed lipid droplets [113]. Furthermore, forced expression of perilipin which leads to the accumulation of lipid droplets within wild type adipocytes becomes markedly impaired in those deficient for caveolin-1 [115].

As well as mediating triglyceride synthesis, caveolins may also function as important modulators of lipolysis (i.e. triglyceride hydrolysis). For example, caveolin-1 has been implicated in facilitating the PKA (protein kinase A) mediated phosphorylation of perilipin – a key step in the lipolytic process [115]. In addition, expression of a dominant-negative truncation mutant of caveolin-3 which specifically associates with the membrane surrounding large lipid droplets, has been shown to inhibit both the catabolism and motility of lipid droplets [112]. Consistent with these findings, caveolin-1 deficient adipocytes exhibit impaired lipolysis in response to β-adrenergic stimulation [115]. Therefore, caveolin proteins may act, potentially through processes initiated with CEMs, to coordinate the cellular machinery required for the synthesis and storage of triglycerides.

As well as facilitating the synthesis and breakdown of triglycerides, CEMs may also regulate the cellular uptake of fatty acids which constitute these neutral lipids. Importantly, disruption of CEMs by β-cyclodextrin treatment or expression of a dominant-negative mutant of caveolin-3 has been shown to suppress oleic acid (C18:1) uptake into 3T3-L1 adipocytes [112]. In the same study, incubation of 3T3-L1 adipocytes with labelled stearate (C18:0) was found to result in the localisation of the long chain saturated fatty acid within caveolin-1 positive detergent resistant fractions. Consistent with this, caveolin-1 possesses a lipid binding site which may be required for trafficking lipid molecules across cell membranes [116]. Furthermore, a subsequent study by Ring and co-workers demonstrated that caveolin-1 is required for promoting the recruitment of fatty acid translocase (FAT/CD36) to the plasma membrane, thereby regulating fatty acid uptake through its ability to modulate the surface availability of key fatty acid transporter proteins [117].

Another potential way that CEMs/caveolins may beneficially impact upon adipocyte function is through attenuating macrophage infiltration into adipose tissue. In the obese state, increased adipose tissue macrophage (ATM) infiltration has been proposed to contribute to insulin resistance and metabolic dysfunction through macrophage-mediated production and secretion of pro-inflammatory cytokines [118-123]. Interestingly, adipose tissue isolated from caveolin-1 or cavin-1 knockout mice has been shown to exhibit greater macrophage infiltration relative to control wild type tissue [45, 124]. Furthermore, a study by Briand and co-workers found that the observed increase in perigenital adipose ATM infiltration in caveolin-1 null mice can be normalised with specific restoration of caveolin-1 within the endothelial cells of these animals [124].

Collectively, these findings indicate that CEMs/caveolins act to regulate a number of different processes, either directly within adipocytes or indirectly through other cell types, which can impact upon adipocyte function. The following section discusses the potential role of caveolins as functional modulators of mitochondria – organelles which function as cellular energy generators.

Caveolins as Modulators of Mitochondrial Function

Mitochondria are responsible for generating most (around 90%) of cellular energy in the form of ATP. These organelles are surrounded by a smooth outer membrane and an inner membrane folded to produce layers, known as cristae which are studded with various proteins required for ATP synthesis. The leaflets of the inner membrane are exposed to the gel-like mitochondrial matrix harbouring enzymes that catalyse a series of reactions resulting in the generation of reduced cofactors. It is the transfer of electrons from these reduced cofactors to a series of protein complexes known as the electron transport chain which drives the activity of ATP synthase, the enzyme responsible for generating ATP from ADP and inorganic phosphate. Collectively, the process of generating ATP by the electron transport chain is termed oxidative phosphorylation (OXPHOS).

Although caveolins are known to play a key role in the formation of functional caveolae at the plasma membrane, there is evidence to suggest that these proteins may also be targeted to mitochondria. For example, several studies involving analysis by immunoblotting, immunoprecipitation and/or electron microscopy (immunogold labelling) have detected caveolin-1, albeit at low levels, within the mitochondria of several cell types [125]. Although the mechanism by which caveolin-1 becomes targeted to mitochondria remains unclear, it most likely involves the transport of a soluble form of caveolin-1 as part of a lipid complex (i.e. contained within lipid droplets/vesicles) [125].

Importantly, loss of caveolin-1 expression/function has been shown to impair mitochondrial function in several different ways. Firstly, a study by Bosch and colleagues first demonstrated that caveolin-1 deficiency promotes the accumulation of cholesterol within mitochondrial membranes [126]. Crucially, this increase in cholesterol content was associated with reduced membrane fluidity and impaired efficiency of electron transport. Moreover, cholesterol depleting caveolin-1 deficient mitochondria was found to restore coupling between mitochondrial respiration and ATP production [126]. Therefore, one way that caveolin isoforms may act to maintain mitochondrial integrity and function is through their ability to modulate the cholesterol content of membranes which constitute this organelle.

Imaging based studies have also revealed dramatic morphological alterations in mitochondria within brown adipocytes which lack caveolin-1 [127]. In particular, caveolin-1 deficient adipocytes have been shown to contain a significantly reduced number of mitochondria, which themselves display increased swelling of the mitochondrial matrix [127]. Furthermore, a study by Asterholm and co-workers showed that several genes considered important for mitochondrial function, including those implicated in oxidative phosphorylation, are significantly downregulated in adipose tissue from caveolin-1 deficient mice compared to wild type control animals [128]. Interestingly, authors of the same study were also able to demonstrate that mouse embryonic fibroblasts (MEFs) lacking caveolin-1 display a preference for glycolysis when maintained under glucose-free conditions, whilst wild type MEFs show increased oxygen consumption – indicative of enhanced mitochondrial oxidative capacity. In accordance with these observations, targeted gene transfer of functional caveolin-1 to mitochondria has been shown to enhance respiratory function in cardiac myocytes [129].

Further work will be required to establish whether caveolae/caveolins present within the plasma membrane are also able to modulate processes that support mitochondrial formation and/or activity. In addition, it remains unclear exactly what factors may trigger the translocation of caveolin proteins to mitochondria. Importantly, strategies designed to enhance trafficking of caveolins to mitochondria may not only facilitate in improving mitochondrial oxidative capacity (e.g. fatty acid β-oxidation), but also enhance those processes that depend upon the ATP generating capacity of these organelles (e.g. maintenance of transmembrane ion gradients and protein synthesis). Indeed, resolving these issues may allow a better understanding of the extent to which caveolae/caveolins contribute towards mitochondrial dysfunction and how this is linked to the development of metabolic disorders such as obesity, insulin resistance and diabetes [130-132].

The Role of Caveolae in Human Disease - Involvement of Genetic and Non-Genetic Factors

Several clinical studies suggest that caveolar dysfunction may underlie the pathogenesis of a number of human diseases – some of which are associated with metabolic impairments. For example, individuals harbouring mutations in the CAV-1 gene have been found to display varying extents of dyslipidemia (in the form of lipodystrophy) and insulin resistance [133-135]. In one particular study examining hypertensive subjects, it was found that individuals carrying single nucleotide polymorphisms in the CAV1 gene were more likely to exhibit increased fasting insulin levels and greater insulin resistance [134]. In addition, mutations in the gene encoding for cavin-1 have also been linked to the development of lipodystrophy [136].

Non-genetic factors may also play an important role in modulating caveolar function in human tissues. For example, both caveolin-1 and caveolin-3 protein content were found to be markedly elevated in EDL muscle of rats following resistance exercise [137]. Importantly, this upregulation of caveolin-1 and caveolin-3 protein expression coincided with increased insulin receptor β subunit protein content and enhanced phosphorylation of PKB/Akt – indicative of enhanced insulin sensitivity. Indeed, such increases in the expression of caveolin proteins may at least partly underlie the beneficial metabolic effects associated with physical exercise.

In accordance with the idea that caveolin isoforms are sensitive to their lipid environment, several studies suggest that certain dietary fatty acids may impact upon caveolar function. For example, omega-3 polyunsaturated fatty acids (PUFAs) such as eicopentaenoic acid (C20:5) have been shown to promote alterations in the lipid composition of T cell lipid rafts, in particular increasing their unsaturated fatty acyl content [138-140]. Importantly, increased dietary intake of PUFAs has been reported to improve a number of clinical conditions including diabetes, cardiovascular disease and inflammatory disorders [140]. In accordance with this, omega-3 PUFAs have been shown to stimulate mitochondrial biogenesis and fatty acid β-oxidation, reduce systemic inflammation and enhance insulin sensitivity [141-145]. Therefore, it is possible that such beneficial metabolic responses conveyed by PUFAs may be partly due to their ability to alter the lipid and/or protein composition of lipid rafts/caveolae and/or act upon components localised within these membrane microdomains. In contrast, diets enriched with saturated fats may cause impaired insulin sensitivity through promoting deleterious changes in caveolar composition, for example by suppressing caveolin expression [146].

It should also be highlighted that improved metabolic profiles conveyed by pharmacological agents that have either previously been or are currently used in the clinical treatment of type II diabetes, may also be caveolar dependent. For example, thiazolidinedione PPAR (peroxisome proliferator-activated receptor) ligands such as rosiglitazone which act as insulin sensitizers, have been reported to upregulate caveolin mRNA and protein abundance in macrophages and carcinoma cells [147, 148]. Therefore, it is possible that these PPAR ligands may act to enhance insulin sensitivity through inducing similar changes in caveolar components within key insulin target tissues such as skeletal muscle and adipose tissue.

In addition, counteracting the insulin desensitising effects of caveolar lipids such as ceramide or GM3, for example through suppressing their synthesis or inhibiting the activity of downstream effector kinases (e.g. atypical PKC isoforms in the case of ceramide), may provide a useful therapeutic strategy for combating obesity-induced insulin resistance. Indeed, impairing ceramide or GM3 synthesis has previously been reported to enhance insulin sensitivity in rodent models of diet-induced obesity [90, 149, 150]. Alternatively, targeted gene transfer of caveolins into key metabolic tissues may also provide a means of improving insulin sensitivity and metabolic capacity. Whether these anti-diabetic therapeutic strategies would be effective in treating obesity-related metabolic disorders in humans remains to be explored.

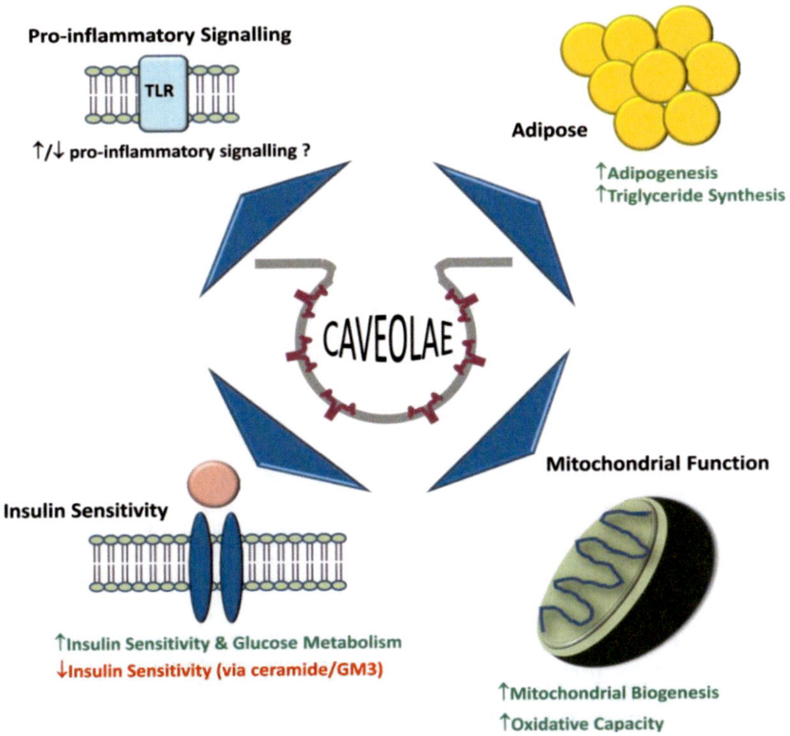

Figure 6. Summary of caveolae dependent regulation of energy homeostasis. Schematic diagram summarises how caveolar microdomains may regulate various processes which are important for energy homeostasis.

Conclusion

A number of studies have served to highlight the key role that lipid rafts/caveolae play in the modulation of energy homeostasis. Growing evidence now suggests that these membrane microdomains act as crucial signalling foci for the initiation and propagation of numerous metabolic processes including insulin signal transduction, nutrient transport, pro-inflammatory signalling, lipid metabolism and mitochondrial function (see Figure 6). However, further work will be required to establish whether obesity-associated factors can promote changes in caveolar composition and function, and importantly, how such alterations may impact upon metabolic signalling. Importantly, as this research area progresses, it promises to unravel novel mechanistic insights into the development of insulin resistance and metabolic dysfunction, which in turn could provide a basis for the identification of new therapeutic targets in the treatment of obesity-induced metabolic disorders.

Acknowledgments

Research in the authors' laboratory is supported by the BBSRC, Scottish Government's Rural and Environment Science and Analytical Services Division (RESAS) and Diabetes UK.

References

[1] Foster, L.J., and Chan, Q.W. (2007) *Subcell Biochem* 43, 35-47

[2] Lucero, H.A., and Robbins, P.W. (2004) *Arch Biochem Biophys* 426, 208-224

[3] Yamada, E. (1955) 1,445-458

[4] Fra, A.M., Williamson, E., Simons, K., and Parton, R.G. (1995) *Proc Natl Acad Sci U S A* 92, 8655-8659

[5] Galbiati, F., Volonte, D., Engelman, J.A., Watanabe, G., Burk, R., Pestell, R.G., and Lisanti, M.P. (1998) *EMBO J* 17, 6633-6648

[6] Drab, M., Verkade, P., Elger, M., Kasper, M., Lohn, M., Lauterbach, B., Menne, J., Lindschau, C., Mende, F., Luft, F.C., Schedl, A., Haller, H., and Kurzchalia, T.V. (2001) *Science* 293, 2449-2452

[7] Galbiati, F., Razani, B., and Lisanti, M.P. (2001) *Trends Mol Med* 7,435-441

[8] Park, D.S., Lee, H., Frank, P.G., Razani, B., Nguyen, A.V., Parlow, A.F., Russell, R.G., Hulit, J., Pestell, R.G., and Lisanti, M.P. (2002) *Mol Biol Cell* 13, 3416-3430

[9] Razani, B., Combs, T.P., Wang, X.B., Frank, P.G., Park, D.S., Russell, R.G., Li, M., Tang, B., Jelicks, L.A., Scherer, P.E., and Lisanti, M.P. (2002) *J Biol Chem* 277, 8635-8647

[10] Woodman, S.E., Park, D.S., Cohen, A.W., Cheung, M.W., Chandra, M., Shirani, J., Tang, B., Jelicks, L.A., Kitsis, R.N., Christ, G.J., Factor, S.M., Tanowitz, H.B., and Lisanti, M.P. (2002) *J Biol Chem* 277, 38988-38997

[11] Scherer, P.E., Tang, Z., Chun, M., Sargiacomo, M., Lodish, H.F., and Lisanti, M.P. (1995) *J Biol Chem* 270, 16395-16401

[12] Kogo, H., and Fujimoto, T. (2000) *FEBS Lett* 465, 119-123

[13] Kogo, H., Aiba, T., and Fujimoto, T. (2004) *J Biol* 279, 25574-25581

[14] Williams, T.M., and Lisanti, M.P. (2004) *Genome Biol* 5:214

[15] Monier, S., Parton, R.G., Vogel, F., Behlke, J., Henske, A., and Kurzchalia, T.V. (1995) *Mol Biol Cell* 6:911-927

[16] Gonzalez-Munoz, E., Lopez-Iglesias, C., Calvo, M., Palacin, M., Zorzano, A., and Camps, M. (2009) 150, 3493-3502

[17] Sargiacomo, M., Scherer, P.E., Tang, Z., Kubler, E., Song, K.S., Sanders, M.C., and Lisanti, M.P. (1995) *Proc Natl Acad Sci U S A* 92, 9407-9411

[18] Vassilopoulos, S., Oddoux, S., Groh, S., Cacheux, M., Faure, J., Brocard, J., Campbell, K.P., and Marty, I. (2010) *Biochemistry* 49, 6130-6135

[19] Hagiwara, Y., Sasaoka, T., Araishi, K., Imamura, M., Yorifuji, H., Nonaka, I., Ozawa, E., and Kikuchi, T. (2000) *Hum Mol Genet* 9, 3047-3054

[20] Murata, M., Peranen, J., Schreiner, R., Wieland, F., Kurzchalia, T.V., and Simons, K. (1995) *Proc Natl Acad Sci U S A* 92, 10339-10343

[21] Hailstones, D., Sleer, L.S., Parton, R.G., and Stanley, K.K. (1998) *J Lipid Res* 39, 369-379

[22] Dreja, K., Voldstedlund, M., Vinten, J., Tranum-Jensen, J., Hellstrand, P., and Sward, K. (2002) *Arterioscler Thromb Vasc Biol* 22, 1267-1272

[23] Hajduch, E., Turban, S., Le Liepvre, X., Le Lay, S., Lipina, C., Dimopoulos, N., Dugail, I., and Hundal, H.S. (2008) *Biochem J* 410, 369-379

[24] Sommer, B., Montano, L.M., Carbajal, V., Flores-Soto, E., Ortega, A., Ramirez-Oseguera, R., Irles, C., El-Yazbi, A.F., Cho, W.J., and Daniel, E.E. (2009) *Can J Physiol Pharmacol* 87, 180-195

[25] Anderson, H.A., Chen, Y., and Norkin, L.C. (1996) *Mol Biol Cell* 7, 1825-1834

[26] Orlandi, P.A., and Fishman, P.H. (1998) *J Cell Biol* 141, 905-915

[27] Xiang, Y., Rybin, V.O., Steinberg, S.F., and Kobilka, B. (2002) *J Biol Chem* 277, 34280-34286

[28] Byrne, D.P., Dart, C., and Rigden, D.J. (2012) *PLoS One* 7, e44879

[29] Kurzchalia, T.V., and Parton, R.G. (1999) *Curr Opin Cell Biol* 11, 424-431

[30] Fielding, C.J. (2001) *Curr Opin Lipidol* 12, 281-287

[31] Sowa, G. (2000) *Front Physiol* 2:120

[32] Uittenbogaard, A., and Smart, E.J. (2000) *J Biol Chem* 275, 25595-25599

[33] Lee, H., Woodman, S.E., Engelman, J.A., Volonte, D., Galbiati, F., Kaufman, H.L., Lublin, D.M., and Lisanti, M.P. (2001) *J Biol Chem* 276, 35150-35158

[34] Nomura, R., and Fujimoto, T. (1999) *Mol Biol Cell* 10, 975-986

[35] Labrecque, L., Nyalendo, C., Langlois, S., Durocher, Y., Roghi, C., Murphy, G., Gingras, D., and Beliveau, R. (2004) *J Biol Chem* 279, 52132-52140

[36] Urra, H., Torres, V.A., Ortiz, R.J., Lobos, L., Diaz, M.I., Diaz, N., Hartel, S., Leyton, L., and Quest, A.F. (2012) *PLoS One* 7, e33085

[37] Liu, L., and Pilch, P.F. (2008) *J Biol Chem* 283, 4314-4322

[38] Hill, M.M., Bastiani, M., Luetterforst, R., Kirkham, M., Kirkham, A., Nixon, S.J., Walser, P., Abankwa, D., Oorschot, V.M., Martin, S., Hancock, J.F., and Parton, R.G. (2008) *Cell* 132, 113-124

[39] Hansen, C.G., Bright, N.A., Howard, G., and Nichols, B.J. (2009) *Nat Cell Biol* 11, 807-814

[40] Chidlow, J.H. Jr., and Sessa, W.C. (2010) *Cardiovasc Res* 86, 219-225

[41] McMahon, K.A., Zajicek, H., Li, W.P., Peyton, M.J., Minna, J.D., Hernandez, V.J., Luby-Phelps, K., and Anderson, R.G. (2009) *EMBO J* 28, 1001-1015

[42] Tagawa, M., Ueyama, T., Ogata, T., Takehara, N., Nakajima, N., Isodono, K., Asada, S., Takahashi, T., Matsubara, H., and Oh, H. (2008) *Am J Physiol Cell Physiol* 295, C490-498

[43] Bastiani, M., Liu, L., Hill, M.M., Jedrychowski, M.P., Nixon, S.J., Lo, H.P., Abankwa, D., Luetterforst, R., Fernandez-Rojo, M., Breen, M.R., Gygi, S.P., Vinten, J., Walser, P.J., North, K.N., Hancock, J.F., Pilch, P.F., and Parton, R.G. (2009) *J Cell Biol* 185, 1259-1273

[44] Aung, C.S., Hill, M.M., Bastiani, M., Parton, R.G., and Parat, M.O. (2011) *Eur J Cell Biol* 90, 136-142

[45] Govender, P., Romero, F., Shah, D., Paez, J., Ding, S.Y., Liu, L., Gower, A., Baez, E., Aly, S.S., Pilch, P., and Summer, R. (2013) *PLoS One* 8, e62045

[46] Volonte, D., Galbiati, F., Li, S., Nishiyama, K., Okamoto, T., and Lisanti, M.P. (1999) *J Biol Chem* 274, 12702-12709

[47] Fecchi, K., Volonte, D., Hezel, M.P., Schmeck, K., and Galbiati, F. (2006) *FASEB J* 20, 705-707

[48] Banning, A., Tomasovic, A., and Tikkanen, R. (2011) *Curr Protein Pept Sci* 12, 725-735

[49] Amaddii, M., Meister, M., Banning, A., Tomasovic, A., Mooz, J., Rajalingam, K., and Tikkanen, R. (2012) *J Biol Chem* 87, 7265-7278

[50] Schuck, S., Honsho, M., Ekroos, K., Shevchenko, A., and Simons, K. (2003) *Proc Natl Acad Sci U S A* 100, 5795-5800

[51] Williamson, R., Thompson, A.J., Abu, M., Hye, A., Usardi, A., Lynham, S., Anderton, B.H., and Hanger, D.P. (2010) *BMC Neurosci* 11, 120

[52] Hemmings, B.A., and Restuccia, D.F. (2012) *Cold Spring Harb Perspect Biol* 4, a011189

[53] Milburn, C.C., Deak, M., Kelly, S.M., Price, N.C., Alessi, D.R., and Van Aalten, D.M. (2003) *Biochem J* 375, 531-538

[54] Sarbassov, D.D., Guertin, D.A., Ali, S.M., and Sabatini, D.M. (2005) *Science* 307, 1098-1101

[55] Bayascas, J.R. (2008) *Cell Cycle* 7, 2978-2982

[56] Miinea, C.P., Sano, H., Kane, S., Sano, E., Fukuda, M., Peranen, J., Lane, W.S., and Lienhard, G.E. (2005) *Biochem J* 391, 87-93

[57] Cross, D.A., Alessi, D.R., Cohen, P., Andjelkovich, M., and Hemmings, B.A. (1995) *Nature* 378, 785-789

[58] Gustavsson, J., Parpal, S., Karlsson, M., Ramsing, C., Thorn, H., Borg, M., Lindroth, M., Peterson, K.H., Magnusson, K.E., and Stralfors, P. (1993) *FASEB J* 13, 1961-1971

[59] Karlsson, M., Thorn, H., Danielsson, A., Stenkula, K.G., Ost, A., Gustavsson, J., Nystrom, F.H., and Stralfors, P. (2004) *Eur J Biochem* 271, 2471-2479

[60] Smythe, G.M., and Rando, T.A. (2006) *Exp Cell Res* 312, 2816-2825

[61] Foti, M., Porcheron, G., Fournier, M., Maeder, C., and Carpentier, J.L. (2007) *Proc Natl Acad Sci U S A* 104, 1242-1247

[62] Nystrom, F.H., Chen, H., Cong, L.N., Li, Y., and Quon, M.J. (1999) *Mol Endocrinol* 13, 2013-2024

[63] Oh, Y.S., Cho, K.A., Ryu, S.J., Khil, L.Y., Jun, H.S., Yoon, J.W., and Park, S.C. (2006) *J Cell Biochem* 99, 747-758

[64] Kabayama, K., Sato, T., Saito, K., Loberto, N., Prinetti, A., Sonnino, S., Kinjo, M., Igarashi, Y., and Inokuchi, J. (2007) *Proc Natl Acad Sci U S A* 104, 13678-13683

[65] Schubert, W., Frank, P.G., Razani, B., Park, D.S., Chow, C.W., and Lisanti, M.P. (2001) *J Biol Chem* 276, 48619-48622

[66] Razani, B., Engelman, J.A., Wang, X.B., Schubert, W., Zhang, X.L., Marks, C.B., Macaluso, F., Russell, R.G., Li, M., Pestell, R.G., Di Vizio,

D., Hou, H. Jr., Kneitz, B., Lagaud, G., Christ, G.J., Edelmann, W., and Lisanti, M.P. (2001) *J Biol Chem* 276, 38121-38138

[67] Cohen, A.W., Razani, B., Wang, X.B., Combs, T.P., Williams, T.M., Scherer, P.E., and Lisanti, M.P. (2003) *Am J Physiol Cell Physiol* 285, C222-235

[68] Capozza, F., Combs, T.P., Cohen, A.W., Cho, Y.R., Park, S.Y., Schubert, W., Williams, T.M., Brasaemle, D.L., Jelicks, L.A., Scherer, P.E., Kim, J.K., and Lisanti, M.P. (2005) *Am J Physiol Cell Physiol* 288, C1317-1331

[69] Oshikawa, J., Otsu, K., Toya, Y., Tsunematsu, T., Hankins, R., Kawabe, J., Minamisawa, S., Umemura, S., Hagiwara, Y., and Ishikawa, Y. (2004) *Proc Natl Acad Sci U S A* 101, 12670-12675

[70] Otsu, K., Toya, Y., Oshikawa, J., Kurotani, R., Yazawa, T., Sato, M., Yokoyama, U., Umemura, S., Minamisawa, S., Okumura, S., and Ishikawa, Y. (2010) *Am J Physiol Cell Physiol* 298, C450-456

[71] Trajkovski, M., Hausser, J., Soutschek, J., Bhat, B., Akin, A., Zavolan, M., Heim, M.H., and Stoffel, M. (2011) *Nature* 474, 649-653

[72] Chen, J., Capozza, F., Wu, A., Deangelis, T., Sun, H., Lisanti, M., and Baserga, R. (2008) *J Cell Physiol* 217, 281-289

[73] Parpal, S., Karlsson, M., Thorn, H., and Stralfors, P. (2001) *J Biol Chem* 276, 9670-9678

[74] Sanchez-Wandelmer, J., Davalos, A., Herrera, E., Giera, M., Cano, S., de la Pena, G., Lasuncion, M.A., and Busto, R. (2009) *Biochim Biophys Acta* 1788, 1731-1739

[75] Fagerholm, S., Ortegren, U., Karlsson, M., Ruishalme, I., and Stralfors, P. (2009) *PLoS One* 4, e5985

[76] Scherer, P.E., Lisanti, M.P., Baldini, G., Sargiacomo, M., Mastick, C.C., and Lodish, H.F.(1994) *J Cell Biol* 127, 1233-1243

[77] Gustavsson, J., Parpal, S., and Stralfors, P. (1996) *Mol Med* 2, 367-372

[78] Ortegren, U., Yin, L., Ost, A., Karlsson, H., Nystrom, F.H., and Stralfors, P. (2006) *FEBS J* 273, 3381-3392

[79] Watson, R.T., and Pessin, J.E. (2001) *Exp Cell Res* 271, 75-83

[80] Chiang, S.H., Baumann, C.A., Kanzaki, M., Thurmond, D.C., Watson, R.T., Neudauer, C.L., Macara, I.G., Pessin, J.E., and Saltiel, A.R. *Nature* 410, 944-948

[81] Kwon, H., and Pak, Y. (2010) *Biochem Biophys Res Commun* 391, 49-55

[82] Kwon, H., Lee, J., Jeong, K., Jang, D., and Pak, Y. (2013) *Biochim Biophys Acta* 1833, 2176-2189

[83] Schmitz-Peiffer, C., Craig, D.L., and Biden, T.J. (1999) *J Biol Chem* 274, 24202-24210

[84] Powell, D.J., Turban, S., Gray, A., Hajduch, E., and Hundal, H.S. (2004) *Biochem J* 382, 619-629

[85] Blouin, C.M., Prado, C., Takane, K.K., Lasnier, F., Garcia-Ocana, A., Ferre, P., Dugail, I., and Hajduch, E. (2010) *Diabetes* 59, 600-610

[86] Xia, H., Khalil, W., Kahm, J., Jessurun, J., Kleidon, J., and Henke, C.A. (2010) *Am J Pathol* 176, 2626-2637

[87] Lee, H., Xie, L., Luo, Y., Lee, S.Y., Lawrence, D.S., Wang, X.B., Sotgia, F., Lisanti, M.P., and Zhang, Z.Y. (2006) *Biochemistry* 45, 234-240

[88] Goldstein, B.J., Bittner-Kowalczyk, A., White, M.F., and Harbeck, M. (2000) *J Biol Chem* 275, 4283-4289

[89] Parvaneh, L., Meshkani, R., Bakhtiyari, S., Mohammadtaghvaie, N., Gorganifiruzjaee, S., Taheripak, G., Golestani, A., Foruzandeh, M., Larijani, B., and Taghikhani, M. (2010) *Biochem Biophys Res Commun* 396, 467-471

[90] Yamashita, T., Hashiramoto, A., Haluzik, M., Mizukami, H., Beck, S., Norton, A., Kono, M., Tsuji, S., Daniotti, J.L., Werth, N., Sandhoff, R., Sandhoff, K., and Proia, R.L. (2003) *Proc Natl Acad Sci U S A* 100, 3445-3449

[91] van Eijk, M., Aten, J., Bijl, N., Ottenhoff, R., van Roomen, C.P., Dubbelhuis, P.F., Seeman, I., Ghauharali-van der Vlugt, K., Overkleeft, H.S., Arbeeny, C., Groen, A.K., and Aerts, J.M. (2009) *PLoS One* 4, e4723

[92] Sekimoto, J., Kabayama, K., Gohara, K., and Inokuchi, J. (2012) *FEBS Lett* 586, 191-195

[93] Li, Q., and Verma, I.M. (2002) NF-kappaB regulation in the immune system. *Nat Rev Immunol* 2, 725-734

[94] Karin, M., and Ben-Neriah, Y. (2000) *Annu Rev Immunol* 18, 621-663

[95] Hayden, M.S., and Ghosh, S. (2004) *Genes Dev* 18, 2195-2224

[96] Ajuwon, K.M., and Spurlock, M.E. (2005) *J Nutr* 135, 1841-1846

[97] . Bradley, R.L., Fisher, F.F., and Maratos-Flier, E. (2008) *Obesity (Silver Spring)* 16:938-944

[98] Green, C.J., Pedersen, M., Pedersen, B.K., and Scheele, C. (2011) *Diabetes* 60, 2810-2819

[99] Fuentes, E., Fuentes, F., Vilahur, G., Badimon, L., and Palomo, I. (2013) *Mediators Inflamm* 2013, 136584

[100] Aguirre, V., Uchida, T., Yenush, L., Davis, R., and White, M.F. (2000) *J Biol Chem* 275, 9047-9054
[101] Hirosumi, J., Tuncman, G., Chang, L., Gorgun, C.Z., Uysal, K.T., Maeda, K., Karin, M., and Hotamisligil, G.S. (2002) *Nature* 420, 333-336
[102] Rui, L., Aguirre, V., Kim, J.K., Shulman, G.I., Lee, A., Corbould, A., Dunaif, A., and White, M.F. (2001) *J Clin Invest* 107, 181-189
[103] Uysal, K.T., Wiesbrock, S.M., Marino, M.W., and Hotamisligil, G.S. (1997) *Nature* 389, 610-614
[104] Arkan, M.C., Hevener, A.L., Greten, F.R., Maeda, S., Li, Z.W., Long, J.M., Wynshaw-Boris, A., Poli, G., Olefsky, J., and Karin, M. (2005) *Nat Med* 2005;11:191-198
[105] Garrean, S., Gao, X.P., Brovkovych, V., Shimizu, J., Zhao, Y.Y., Vogel, S.M., and Malik, A.B. (2006) *J Immunol* 177, 4853-4860
[106] Yuan, K., Huang, C., Fox, J., Gaid, M., Weaver, A., Li ,G., Singh, B.B., Gao, H., and Wu, M. (2011) *J Biol Chem* 286, 21814-21825
[107] Guo, Q., Shen, N., Yuan, K., Li, J., Wu, H., Zeng, Y., Fox, J. 3rd, Bansal, A.K., Singh, B.B., Gao, H., and Wu, M. (2012) *Eur J Immunol* 42,1500-1511
[108] Cheng, A.M., Handa, P., Tateya, S., Schwartz, J., Tang, C., Mitra, P., Oram, J.F., Chait, A., and Kim, F. (2012) *PLoS One* 7, e33917
[109] Wang, X.M., Kim, H.P., Nakahira, K., Ryter, S.W., and Choi, A.M. (2009) *J Immunol* 182:3809-3818
[110] Wang, X.X., Wu, Z., Huang, H.F., Han, C., Zou, W., and Liu, J. (2013) *Asian Pac J Cancer Prev* 14, 2295-2299
[111] Ost, A., Ortegren, U., Gustavsson, J., Nystrom, F.H., and Stralfors, P. (2005) *J Biol Chem* 280, 5-8
[112] Pohl, J., Ring, A., Ehehalt, R., Schulze-Bergkamen, H., Schad, A., Verkade, P., and Stremmel, W. (2004) *Biochemistry* 43, 4179-4187
[113] Le Lay, S., Hajduch, E., Lindsay, M.R., Le Liepvre, X., Thiele, C., Ferre, P., Parton, R.G., Kurzchalia, T., Simons, K., and Dugail, I. (2006) *Traffic* 7:549-561
[114] Blouin, C.M., Le Lay, S., Lasnier, F., Dugail, I., and Hajduch, E. (2008) *Biochem Biophys Res Commun* 376, 331-335
[115] . Cohen, A.W., Razani, B., Schubert, W., Williams, T.M., Wang, X.B., Iyengar, P., Brasaemle, D.L., Scherer, P.E., and Lisanti, M.P. (2004) *Diabetes* 53, 1261-1270
[116] Trigatti, B.L., Anderson, R.G., and Gerber, G.E. (1999) *Biochem Biophys Res Commun* 255, 34-39

[117] Ring, A., Le Lay, S., Pohl, J., Verkade, P., and Stremmel, W. (2006) *Biochim Biophys Acta* 1761, 416-423
[118] Weisberg, S.P., Hunter, D., Huber, R., Lemieux, J., Slaymaker, S., Vaddi, K., Charo, I., Leibel, R.L., and Ferrante, A.W. Jr. (2006) *J Clin Invest* 116, 115-124
[119] Feral, C.C., Neels, J.G., Kummer, C., Slepak, M., Olefsky, J.M., and Ginsberg, M.H. (2008) *Diabetes* 57, 1842-1851
[120] Kanda, H., Tateya, S., Tamori, Y., Kotani, K., Hiasa, K., Kitazawa, R., Kitazawa, S., Miyachi, H., Maeda, S., Egashira, K., and Kasuga, M. (2006) *J Clin Invest* 116, 1494-1505
[121] Ortega Martinez de Victoria,. E., Xu, X., Koska, J., Francisco, A.M., Scalise, M., Ferrante, A.W. Jr., and Krakoff, J. (2009) *Diabetes* 58, 385-393
[122] Oh, D.Y., Morinaga, H., Talukdar, S., Bae, E.J., and Olefsky, J.M. (2012) *Diabetes* 61, 346-354
[123] Sorisky, A., Molgat, A.S., and Gagnon, A. (2013) *Adv Nutr* 4, 67-75
[124] Briand, N., Le Lay, S., Sessa, W.C., Ferre, P., and Dugail, I. (2011) *Diabetes* 60, 448-453
[125] Li, W.P., Liu, P., Pilcher, B.K., and Anderson, R.G. (2001) *J Cell Sci* 114, 1397-1408
[126] Bosch, M., Mari, M., Herms, A., Fernandez, A., Fajardo, A., Kassan, A., Giralt, A., Colell, A., Balgoma, D., Barbero, E., Gonzalez-Moreno, E., Matias, N., Tebar, F., Balsinde, J., Camps, M., Enrich, C., Gross, S.P., Garcia-Ruiz, C., Perez-Navarro, E., Fernandez-Checa, J.C., and Pol, A. (2011) *Curr Biol* 21, 681-686
[127] Cohen, A.W., Schubert, W., Brasaemle, D.L., Scherer, P.E., and Lisanti, M.P. (2005) *Diabetes* 54, 679-686
[128] Asterholm, I.W., Mundy, D.I., Weng, J., Anderson, R.G., and Scherer, P.E. (2012) *Cell Metab* 15, 171-185
[129] Fridolfsson, H.N., Kawaraguchi, Y., Ali, S.S., Panneerselvam, M., Niesman, I.R., Finley, J.C., Kellerhals, S.E., Migita, M.Y., Okada, H., Moreno, A.L., Jennings, M., Kidd, M.W., Bonds, J.A., Balijepalli, R.C., Ross, R.S., Patel, P.M., Miyanohara, A., Chen, Q., Lesnefsky, E.J., Head, B.P., Roth, D.M., Insel, P.A., and Patel, H.H. (2012) *FASEB J* 26, 4637-4649
[130] Kim, J.Y., Hickner, R.C., Cortright, R.L., Dohm, G.L., and Houmard, J.A. (2000) *Am J Physiol Endocrinol Metab* 279, E1039-1044
[131] Kelley, D.E., He, J., Menshikova, E.V., and Ritov, V.B. (2002) *Diabetes* 51, 2944-2950

[132] Sparks, L.M., Xie, H., Koza, R.A., Mynatt, R., Hulver, M.W., Bray, G.A., and Smith, S.R. (2005) *Diabetes* 54, 1926-1933
[133] Cao, H., Alston, L., Ruschman, J., and Hegele, R.A. (2008) *Lipids Health Dis* 7, 3
[134] Pojoga, L.H., Underwood, P.C., Goodarzi, M.O., Williams, J.S., Adler, G.K., Jeunemaitre, X., Hopkins, P.N., Raby, B.A., Lasky-Su, J., Sun, B., Cui, J., Guo, X., Taylor, K.D., Chen, Y.D., Xiang, A., Raffel, L.J., Buchanan, T.A., Rotter, J.I., and Williams, G.H. (2011) *J Clin Endocrinol Metab* 96, E1288-1292
[135] Kim, C.A., Delepine, M., Boutet, E., El Mourabit, H., Le Lay, S., Meier, M., Nemani, M., Bridcl, E., Leite, C.C., Bertola, D.R., Semple, R.K., O'Rahilly, S., Dugail, I., Capeau, J., Lathrop, M., and Magre, J. (2008) *J Clin Endocrinol Metab* 93, 1129-1134
[136] Hayashi, Y.K., Matsuda, C., Ogawa, M., Goto, K., Tominaga, K., Mitsuhashi, S., Park, Y.E., Nonaka, I., Hino-Fukuyo, N., Haginoya, K., Sugano, H., and Nishino, I. (2009) *J Clin Invest* 119, 2623-2633
[137] Oh, Y.S., Kim, H.J., Ryu, S.J., Cho, K.A., Park, Y.S., Park, H., Kim, M., Kim, C.K., and Park, S.C. (2007) *Exp Mol Med* 39, 395-401
[138] Stulnig, T.M., Berger, M., Sigmund, T., Raederstorff, D., Stockinger, H., and Waldhausl, W. (1998) *J Cell Biol* 143, 637-644
[139] Stulnig, T.M., Huber, J., Leitinger, N., Imre, E.M., Angelisova, P., Nowotny, P., and Waldhausl, W. (2001) *J Biol Chem* 276, 37335-37340
[140] Ma, D.W., Seo, J., Switzer, K.C., Fan, Y.Y., McMurray, D.N., Lupton, J.R., and Chapkin, R.S. (2004) *J Nutr Biochem* 15, 700-706
[141] Rasic-Milutinovic, Z., Perunicic, G., Pljesa, S., Gluvic, Z., Sobajic, S., Djuric, I., and Ristic, D. (2007) *Ren Fail* 29, 321-329
[142] Ramel, A., Martinez, A., Kiely, M., Morais, G., Bandarra, N.M., and Thorsdottir, I. (2008) *Diabetologia* 51, 1261-1268
[143] Itariu, B.K., Zeyda, M., Hochbrugger, E.E., Neuhofer, A., Prager, G., Schindler, K., Bohdjalian, A., Mascher, D., Vangala, S., Schranz, M., Krebs, M., Bischof, M.G., and Stulnig, T.M. (2012) *Am J Clin Nutr* 96, 1137-1149
[144] Micallef, M.A., and Garg, M.L. (2009) *Atherosclerosis* 204, 476-482
[145] Flachs, P., Ruhl, R., Hensler, M., Janovska, P., Zouhar, P., Kus, V., Macek Jilkova, Z., Papp, E., Kuda, O., Svobodova, M., Rossmeisl, M., Tsenov, G., Mohamed-Ali, V., and Kopecky, J. (2011) *Diabetologia* 54, 2626-2638
[146] Knowles, C.J., Cebova, M., and Pinz, I.M. (2013) *PLoS One* 8, e61369

[147] Burgermeister, E., Tencer, L., and Liscovitch, M. (2003) *Oncogene* 22, 3888-3900

[148] Llaverias, G., Vazquez-Carrera, M., Sanchez, R.M., Noe, V., Ciudad, C.J., Laguna, J.C., and Alegret, M. (2004) *J Lipid Res* 45, 2015-2024

[149] Yang, G., Badeanlou, L., Bielawski, J., Roberts, A.J., Hannun, Y.A., and Samad, F. (2009) *Am J Physiol Endocrinol Metab* 297, E211-224

[150] Ussher, J.R., Koves, T.R., Cadete, V.J., Zhang, L., Jaswal, J.S., Swyrd, S.J., Lopaschuk, D.G., Proctor, S.D., Keung, W., Muoio, D.M., and Lopaschuk, G.D. (2010) *Diabetes* 59, 2453-2464

[151] Besse, I.M., Mitchell, C.C., Hund, T.J., and Shibata, E.F. (2011) *Front Physiol* 2, 87

[152] de Oliveira, V.E., Almeida, E.W., Castro, H.V., Edwards, H.G., Dos Santos, H.F., and de Oliveira, L.F. (2011) *J Phys Chem A* 115, 8511-8519

[153] Huang, S., Rutkowsky, J.M., Snodgrass, R.G., Ono-Moore, K.D., Schneider, D.A., Newman, J.W., Adams, S.H., and Hwang, D.H. (2012) *J Lipid Res* 53, 2002-2013

[154] Oka, N., Yamamoto, M., Schwencke, C., Kawabe, J., Ebina, T., Ohno, S., Couet, J., Lisanti, M.P., and Ishikawa, Y. (1998) *J Biol Chem* 272, 33416-33421

[155] Yamamoto, M., Toya, Y., Schwencke, C., Lisanti, M.P., Myers, M.G. Jr., and Ishikawa, Y. (1998) *J Biol Chem* 273, 26962-26968

[156] Jablonski, E.M., and Hughes, F.M. Jr. (2006) *Biol Cell* 98, 33-42

[157] Taira, J., and Higashimoto, Y. (2013) *Biochem Biophys Res Commun* 431, 724-728

[158] Di Vizio, D., Adam, R.M., Kim, J., Kim, R., Sotgia, F., Williams, T., Demichelis, F., Solomon, K.R., Loda, M., Rubin, M.A., Lisanti, M.P., and Freeman, M.R. (2008) *Cell Cycle* 7, 2257-2267

[159] Smith, R.M., Harada, S., Smith, J.A., Zhang, S., and Jarett, L. (1998) *Cell Signal* 10, 355-362

[160] Chun, J., Kwon, T., Lee, E.J., Hyun, S., Hong, S.K., and Kang, S.S. (2005) *Biochem Biophys Res Commun* 326, 136-146

[161] Caselli, A., Mazzinghi, B., Camici, G., Manao, G., and Ramponi, G. (2002) *Biochem Biophys Res Commun* 296, 692-697

In: Lipid Rafts
Editor: Dan Sillence

ISBN: 978-1-62948-961-2

Chapter V

Maintaining a Healthy Immune Response by Balancing the Lipid Content of Lipid Raft Microdomains

***Georgia McDonald*[1], *Panagiotis S. Kabouridis*[2] and *Elizabeth C. Jury*[1,*]**

[1]Centre for Rheumatology Research, Department of Medicine, University College London, Rayne Building, London, UK
[2]William Harvey Research Institute, Barts and The London School of Medicine and Dentistry, Queen Mary University of London, Charterhouse Square, London, UK

Abstract

Lipid rafts are cholesterol and glycosphingolipid (GSL) enriched microdomains in the plasma membrane that form areas of relative liquid-

* Corresponding Author: Elizabeth Jury PhD, Centre for Rheumatology Research, University College London, Rayne Building, 5 University Street, London W1CE 6JF, UK. Telephone: 44(0)20 3108 2154; Fax: 44(0)20 3108 2152. E-mail: e.jury@ucl.ac.uk.
The authors declare they have no conflict of interest in relation to this publication. ECJ is supported by Arthritis Research UK and Lupus UK.

order compared to the surrounding disordered membrane. Cholesterol and GSLs are not only intrinsic components of the lipid raft structure but, because they compartmentalize signaling proteins and provide an environment facilitating immunoreceptor signal transduction, they are integral to the regulation of immune cell function; this includes T and B lymphocyte antigen-receptor-mediated signaling, apoptosis and cell surface protein recycling and endocytosis. Given their importance it is not surprising that the control of cholesterol and GSL biosynthesis is tightly regulated involving interactions with serum lipids associated with lipoproteins. How lipid rafts, and significantly how alterations in the lipids associated with them, influence immune processes is still not fully understood. However, a growing body of evidence suggests that membrane lipid composition and homeostasis can regulate the capacity of immune cells to respond correctly to micro-environmental stimuli. Alterations in such metabolic processes contribute to the immune cell dysfunction in conditions such as atherosclerosis, autoimmunity and cancer. This raises the exciting possibility that targeting membrane lipids could control or alter immune cell activation and may be an important therapeutic mechanism to control aberrant immune responses.

Keywords: Lipid rafts, Cholesterol, Glycosphingolipids, Liver X Receptor (LXR)

Introduction

A main issue when considering lipid raft microdomains is their visualization [1] since they are dynamic and of a size too small to resolve using conventional microscopy [2]. Previously, restrictions in imaging technologies and the indirect methods used to assess lipid raft function have caused their existence to be questioned [3, 4]. However, advancements in microscopy and spectroscopy using techniques such as fluorescence resonance energy transfer, single fluorochrome tracking microscopy and photonic force microscopy have aided their elucidation in the membrane of living cells [5-8]. Cellular cholesterol is an important structural component of lipid rafts comprising 20-25% of their total lipid content [9]. Cholesterol preferentially intercalates into the saturated acyl chains of sphingolipids promoting their close packing at the plane of the plasma membrane and bestowing the liquid-ordered properties characteristic of lipid raft domains. [10]. Therefore, cholesterol increases the lateral ordering of lipids, and consequently decreases fluidity of the membrane and reduces permeability of polar molecules [9]. The ability of cell surface proteins to preferentially associate with ordered lipid rafts while others are excluded and diffuse freely in the more disordered membrane [11] is dependent partly on certain protein modifications including myristoylation, palmitoylation and acetylation [12]. Recent work has focused on the relative membrane lipid order in T cell plasma membranes which can be detected using fluorescent membrane probes such as LAURDAN and di-4-ANEPPDHQ (ANE) which partition into both liquid-ordered (raft) and liquid-disordered (non-raft) membranes and detect the environmental difference between the two regions [13]. High membrane order measured by ANE or LAURDAN is required for stable immune synapse (IS) formation and correlates with T cell differentiation. T cells with relatively high plasma membrane lipid order are skewed towards an anti-inflammatory T helper(h)-2 phenotype whereas T cells with lower plasma membrane lipid order exhibit a more pro-inflammatory Th1 phenotype [14-17]. High membrane lipid order also correlates with increased proliferative ability and formation of stable IS between T cells and antigen presenting cells (APC) [14]. Table 1 summarises the characteristics of CD4+ T cells with relative high and low membrane lipid order detected by ANE labelling in healthy human donors.

Cholera toxin B (CTB), which binds to the glycosphingolipid GM1, has been used extensively as a surrogate marker for lipid rafts. However, there are many other GSL species in the plasma membrane which participate in lipid raft functions. The cellular expression of GSLs is established by a combination of several factors including the level and activity of enzymes involved in their metabolic biosynthesis, the relative rate of GSL endocytosis and exocytosis, plasma membrane turnover as determined by cell proliferation status, and external stimuli eliciting GSL degradation or biosynthesis [18, 19].

The study of lysosomal storage diseases (LSDs) such as Niemann Pick type C (NPC), Fabry's and Gaucher's diseases has shown that alterations in lipid metabolism can lead to a plethora of clinical manifestations. LSDs result from mutations in single proteins involved in either lipid trafficking (NPC2 protein) or metabolism (α-galactosidase A, Gaucher; β-glucocerebrosidase) respectively [20]. The implications of aberrant GSL metabolism are also highlighted by experimental mutant GSL synthase mouse and invertebrate models; the majority of which result in either embryonic lethality or severe phenotypical abnormalities [21]. Whether certain aspects of the pathology observed in these diseases are due to lipid raft alterations is an intriguing, but as yet poorly explored, question.

Table I. Characteristics of High, Intermediate and low membrane lipid order CD4+T cells

Membrane order:	High	Intermediate	Low
Proliferation	+++	++	-
IL-2 production	+++	++	ND[1]
IL-4 production	+++	+	ND
IFN-γ production	+	+++	ND
Apoptosis	-	-	+
Associated membrane lipids: GSL (CTB)	+	++	+++
Cholesterol (filipin)	+++	+++	++

[1]ND=not detected.

Thus, very little is known about how differential expression of either cholesterol or GSLs affects immune cell function and whether defects in the regulation of membrane lipids can influence the pathogenesis of a wider range of human disease beyond LSDs.

Lipid rafts and Immune Cell Signalling

The multifaceted process of lymphocyte antigen receptor generation is integral for the development of a diverse repertoire that is lacking in self-reactivity yet it is protective against a number of foreign antigens. The importance of antigen receptor association with lipid rafts to regulate signalling and hence aid the process of development and differentiation has been well established. Initially, the pre-B cell antigen receptor (BCR) and pre–T cell antigen receptor (TCR) are localised within lipid raft microdomains [22, 23]. Conversely, during negative selection of immature T and B cells, the antigen receptors are excluded from rafts, and only form transient and unstable raft associations upon cross-linking [24-26]. It is of interest however that some BCRs are located in lipid rafts in resting B cells, although the functional consequence of such localisation is unknown. It is postulated that they play a role in intracellular signalling leading to cell survival [25].

In unstimulated mature T cells lipid rafts exist transiently and dynamically and are associated with only a few proteins, notably proteins operating proximal to the T cell receptor (TCR), including lymphocyte-specific protein tyrosine kinase (Lck) and linker for activation of T cells (LAT) [24]. T cell activation, via TCR and CD28 engagement, increases the level of raft-associated lipids in the plasma membrane, increasing the size and stability of raft domains. This facilitates the recruitment of important downstream signaling molecules and in particular has been shown to promote lipid raft aggregation at the IS thereby prolonging and amplifying the TCR-generated signal [27-29]. As well as recruiting signaling molecules essential to promote T cell activation, lipid rafts also exclude inhibitory proteins such as programmed cell death protein 1 (PD-1) and the protein tyrosine phosphatase CD45 and CD148 [11, 24, 30-32].

The lipid raft content of B cells is also integral to both BCR and FCεR1 signalling [25]. Inhibitory molecules such as protein tyrosine phosphatase CD45 and CD22 are excluded from lipid rafts in resting B cells [33]. Upon antigen crosslinking, an increased affinity of the BCR to lipid rafts results in its translocation to Src kinase-rich regions that accommodate its phosphorylation and the initiation of the signalling cascade. Highly rapid endocytosis of the BCR upon crosslinking is thought to contribute to an anergic phenotype [34]. Both translocation and endocytosis of the BCR is reliant on the presence of cholesterol in the plasma membrane [34, 35]. Another indication that lipid rafts act as the sites of BCR signalling is recruitment of key downstream signalling molecules to these regions upon activation (Syk, Btk, Vav, SHIP, phospholipase Cγ2, phosphatidylinositol 3-kinase and BLNK) [25]. Lipid rafts also control the surface expression and rate of endocytosis of MHC (major histocompatibility complex) class II molecules on APCs thereby influencing antigen presentation and T cell activation [36-38].

Interestingly, CD1d is a MHC Class I-like molecule that presents lipid antigen to a specialized subset of T cells termed invariant natural killer T (iNKT) cells These cells respond rapidly to lipid antigen presented by CD1d and release a wide range of cytokines and chemokines that trans-activate many other cells of the immune system, thus helping to control the maintenance of immune tolerance and protect against the growth of certain tumours and pathogens, and airway hyper-reactivity [39]. Importantly the increased location of CD1d to lipid rafts in APCs skews the iNKT cell cytokine response from Th1- to Th2 thereby influencing their effect [40].

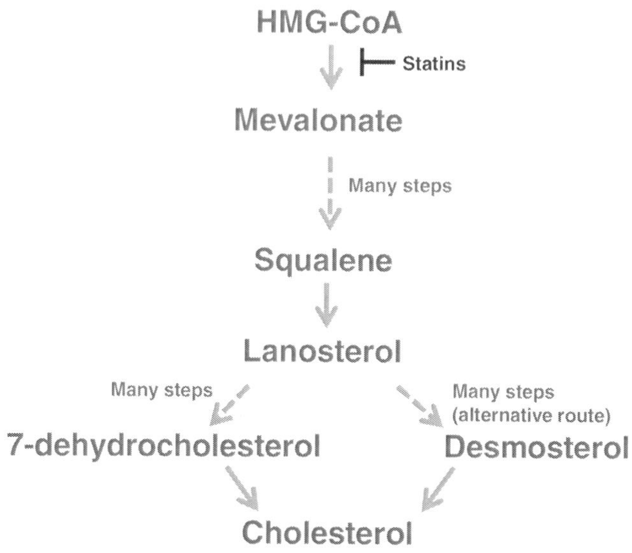

Figure 1. Schematic showing the pathways leading to cholesterol biosynthesis. Membrane incorporation of cholesterol precursors can influence the stability of lipid raft domains. HMG-CoA; 3-hydroxy-3-methylglutaryl-CoA.

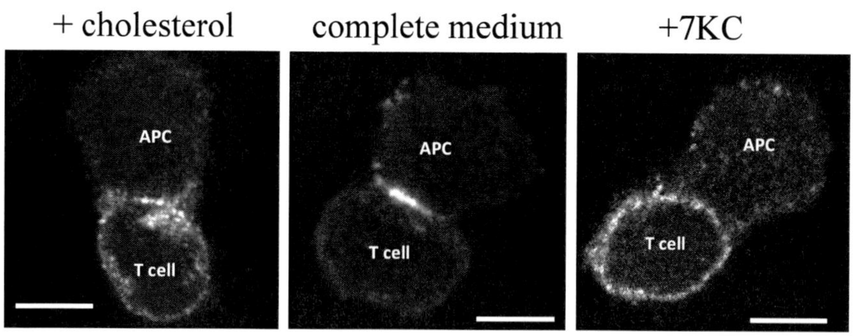

Figure 2. Cholesterol levels influence IS formation. Jurkat T cells were cultured with RPMI medium enriched with cholesterol, 7-ketocholesterol (7KC) or 10% FCS alone (complete medium). T cells were cultured with superantigen-loaded Raji cells as APCs. Jurkat-Raji cell conjugates were fixed, immunostained for anti-CD3 and analyzed by confocal microscopy. The accumulation of CD3 to the IS was disrupted in the presence of excess cholesterol or excess oxysterol. Scale bar =10μM.

How the Lipid Constituents of Rafts Affect Immune Cell Function

Although the importance of membrane protein interaction with lipid rafts during T and B cell activation has been widely investigated, the role of the specific lipid components of rafts is still relatively unexplored [38]. Proteins are selectively incorporated or excluded from lipid raft microdomains depending upon their specific interaction with lipid components or preference for raft structural properties. Recent advances in understanding cholesterol biosynthesis pathways have shown that membrane expression of cholesterol precursors can alter raft structure, stability and function [41, 42]. The cholesterol precursor desmosterol associates less avidly with raft domains than cholesterol itself [41] (Figure 1). In contrast, lathosterol and 7-dehydrocholesterol are better raft stabilizers than cholesterol [42], suggesting that the choice of cholesterol synthesis route may provide a physiological mechanism to modulate raft-dependent functions in cells. Therefore, one consequence of increased membrane cholesterol or an altered balance between cholesterol/cholesterol precursors is that a greater portion of the plasma membrane adopts a liquid-ordered (raft) phase, resulting in altered lateral mobility of receptors and other signaling molecules, including the TCR, thereby influencing signaling molecule interaction at the plasma membrane [43-45].

Our recent work shows that relative membrane lipid order is directly related to the cholesterol content of human $CD4^{+}$T cells [43]. Furthermore, the importance of maintaining a balanced level of plasma membrane cholesterol is demonstrated in Figure 2 where Jurkat T cells loaded with either excess cholesterol or oxidized cholesterol (7-ketocholesterol, 7KC) disrupted the accumulation of TCR-associated CD3 to the IS.

T cells have a distinctive plasma membrane GSL profile or 'fingerprint' that includes monosialoganglioside GM3, the major constituent (72%), a smaller amount of ganglioside GM1, and other minor constituents [46, 47]. Differential GSL expression has considerable importance in a wide variety of T and B cell functions including TCR-mediated signaling [48-50], apoptosis [51] and recycling/endocytosis of membrane signaling and receptor molecules [52]. Cross-linking the raft marker GM1 (using specific binding of CTB-anti-CTB complexes) induces phosphorylation of proximal T cell signaling molecules and calcium mobilization [53] reflecting the interaction of signaling proteins, including src kinases and GPI-anchored proteins, with GSL molecules, either directly or via co-association within lipid rafts [49, 50, 54, 55]. Within T cell rafts themselves, both cholesterol and GSL content is heterogeneous, for example GM1 and GM3 are distributed asymmetrically in polarized T cells [56] and proximal signaling molecules LCK and LAT reside in different raft subsets that fuse upon TCR stimulation, the LCK rafts being more sensitive to cholesterol depletion with methyl-β cyclodextrin (MβCD), while the rafts containing LAT remain intact [57]. Interestingly the interaction of TCR-associated signaling molecules (including Lck and lymphocyte function-associated antigen-1, LFA-1)) with either GM3 or GM1 induces differential recruitment of downstream signaling targets affecting T cell function [58, 59]. Moreover, recent reports show that $CD4^+$ and $CD8^+$T cells exclusively require GM3 (a-series) and GM2 (o-series) GSLs respectively for TCR-mediated activation [54, 60]. Inhibition of TCR signaling has been attributed to inhibition of ZAP-70 microcluster formation by CTLA-4 [61] possibly by strong inhibition of membrane lipid microdomain upregulation [62, 63]. Hence, lipid heterogeneity within lipid rafts contributes to maintaining the appropriate level of membrane protein compartmentalization in resting and activated cells. Thus, changes in GSL and cholesterol expression could have a drastic effect on cellular function [64].

Controlling Lymphocyte Function by Altering the Lipid Content of Rafts

Lipid raft function has been investigated experimentally using methods that alter their cholesterol content including sequestration of cholesterol (with filipin, nystatin or amphotericin), cholesterol depletion using methyl-β-cyclodextrin, inhibition of cholesterol synthesis with HMG-CoA reductase inhibitors (statins) or the addition of oxysterols such as 7-ketocholesterol that reduce membrane order [65]. Disruption of lipid rafts has a number of effects on T cell function; including abrogation of cell-cycle progression and proliferation, reduced sensitivity to Fas-mediated apoptosis, decreased surface expression of co-stimulatory molecules and hence disruption of IS formation [24, 66]. Furthermore changing the balance between cholesterol and oxysterols in the plasma membrane disrupts T cell/APC IS formation, inhibits T cell proliferation and IL-2 production [67, 68].

Attenuation of TCR signaling *in vitro* is also achieved by pharmacological lowering of GSL levels using inhibitors of the biosynthesis pathway, similar results are obtained in GM3 synthase knock-out animals [49, 54, 69]. Reduction of GSL levels inhibits differentiation of $CD4^{+}T$ cells into Th17 cells [69] and reduces cytokine production in activated T cells during viral infection [35]. Interestingly, T cells from patients with the autoimmune disease systemic lupus erythematosis (SLE), have high levels of membrane GM1 and cholesterol associated with the abnormal location of CD45 into lipid rafts and reduced expression of Lck [70, 71]. *In vitro* treatment of these cells with atorvastatin (known to inhibit cholesterol biosynthesis) reduces membrane cholesterol and GSL expression and restores T cell function [72].

Table 2. List of infections and diseases for which lipid rafts and associated proteins are targets

Infections	Diseases
Bacterial:	
Escherichia coli	Alzheimer disease
Mycobacteria tuberculosis	Parkinson disease
Campylobacter jejuni	Polyneuropathies,
Vibrio cholera	demyelinating diseases
Clostridium difficile	
Clostridium tetani	Autoimmune rheumatic diseases:
Salmonella	Systemic lupus erthymatosus
Shigella	Rheumatoid arthritis
Viruses:	Atherosclerosis
Influenza virus	Diabetes
HIV-1	Neoplasia
Measles virus	Asthma and allergic response
Respiratory syncytial cell virus	
Epstein-Barr virus	Lysosomal storage diseases:
	Niemann-Pick disease
Other pathogens:	Tay-Sachs disease,
Plasmodium (malaria)	Fabry's disease,
Trypanosoma (sleeping sickness)	Postsqualene cholesterol biosynthesis
Leishmania	
Prion disease	
Toxoplasma gondii	

Adapted from [120].

The Role of Lipid Rafts in Pathology

It is becoming accepted that lipid rafts and their constituents play a crucial role in the regulation of leukocyte cellular signalling and function. Alterations in plasma membrane lipid content can have a major effect on the nature of the immune response and is likely to play a role in human pathology. Table 2 lists diseases for which lipid rafts and/or their associated proteins can serve as pharmacological targets and some examples are discussed in more detail below.

Infectious Diseases

Lipid rafts facilitate the entry of viruses (including influenza and HIV-1), bacteria and parasites into immune cells. Influenza, which incorporates two integral raft-associated glycoproteins (hemagglutinin and neuraminidase) from host cells, was the first example to be described [73, 74]. HIV-1 likewise integrates host raft-associated lipids and proteins into its envelope, and is dependent on lipid rafts for its life cycle including entry into and budding from T cells. Binding of HIV-1 to T cells causes lipid raft clustering via interactions of CD4 and the raft-associated co-receptors CXCR4 and CCR5 with envelope glycoproteins pg120-gp41 [75]. Both GSLs and cholesterol concentrated in lipid rafts are crucial for the formation of the fusion complex [76, 77]. Importantly, viral translocation can be blocked by disruption of virus-raft association [78].

Furthermore, a group of plasma membrane-associated enzymes, sialidases, are able to modulate membrane GSL content and influence T cell activation during viral infection [35, 79]. Lipid rafts have also been found to activate the type III secretion system (T3SS) required for invasion of host cells by the gram negative bacteria *Salmonella* [33]. Depletion of macrophage membrane cholesterol and blocking of GM1 with CTB inhibited the adhesion and invasion of macrophages by *Trypanosoma cruzi trypomastigotes,* a parasite that causes Sleeping sickness. Furthermore, immunofluorescence imaging revealed colocalization of GM1, flotillin 1 and caveolin within the T. cruzi vacuole [80].

Autoimmune Diseases

T cells from patients with the systemic autoimmune disease SLE are characterised by changes in plasma membrane lipid rafts such as increased levels of cholesterol and GM1 (measured by CTB) [71] and alterations in the overall membrane lipid order [14]. Defects in the distribution of key proximal signalling molecules have been attributed to these changes. The protein tyrosine phosphatase CD45 is key to the conversion of Lck to its active form; increased activation of Lck in T cells from SLE patients was correlated with increased colocalization of CD45 to lipid rafts [64]. In contrast, the negative regulator CTLA-4 is excluded from lipid rafts in SLE T cells, and is therefore unable to control cell signalling and proliferation [81]. All these factors are associated with a reduced threshold for T cell stimulation. Intriguingly, manipulation of membrane lipids by *in vitro* culture with atorvastatin can normalise membrane GM1 expression, phosphorylation of Lck and ERK, and production of IL10 and IL6 in T cells from SLE patients [72]. Furthermore, we have recently observed that $CD4^{+}$T cells from SLE patients displayed an altered profile of lipid raft-associated GSLs, in particular lactosylceramide, Gb3 and GM1, compared to healthy controls [82].

There is evidence for a role of lipid rafts in the pathogenesis of other autoimmune diseases such as rheumatoid arthritis (RA) and multiple sclerosis (MS) [83]. Crucially, certain therapeutic interventions for autoimmune diseases rely on the presence of intact lipid rafts. For example the B cell depletion therapy Rituximab used for the treatment of SLE and RA was shown to induce B cell apoptosis via translocation of CD20 to lipid rafts, an effect that was disrupted upon cholesterol depletion [83].

Cardiovascular Disease

The role of lipid rafts in the pathogenesis of atherosclerosis, a disease characterised by hypercholesteremia and inflammation, has begun to be elucidated. Lipid rafts are suggested to regulate several key events, such as transcytosis of low density lipoproteins (LDLs), immune cell activation and inflammatory response, vascular-wall cell apoptosis, protease activation and plaque stability [84]. Increased lipid raft formation *in vitro* via incubation with LDL/oxidised (ox)LDL induce molecular and immunological changes associated with atherogenesis; translocation of NADPH oxidase, TRPC-1 (transient receptor potential cation channel) and endothelial nitric oxide synthase (eNOS) to lipid rafts which induce pro-inflammatory cytokine release, superoxide production, inhibition of NO secretion and apoptosis. In addition, treatment of atherosclerosis with statins disrupts lipid raft formation and hence pro-atherogenic signalling [84].

Control of Raft-Associated Lipid Expression by Liver X Receptor (LXR)

Since the majority of cellular cholesterol is located within lipid rafts, molecules that control cholesterol biosynthesis could play a role in regulating lipid raft composition. Liver X receptor (LXR) is a master regulator of lipid metabolism. Both LXRα and LXRβ act as a ligand dependent transcription factor located in the nucleus, which upon binding of cholesterol metabolites (oxysterols) form heterodimers with the retinoid X receptor (RXR). Binding of the LXR/RXR complex to LXR-responsive elements (LXREs) elicits transcription of its target genes involved in cellular sterol trafficking [85].

LXR controls cellular cholesterol metabolism via the regulation of downstream genes such as neimann-pick C *(npc1, npc2)*, ATP-binding cassette transporters *(abca1, abcg1)* and the fellow nuclear receptor sterol regulatory element-binding protein *(srebp)* [85]. SREBP functions to promote increase in cellular and hence membrane cholesterol levels via 3-hydroxy-3-methyl-glutaryl-CoA reductase (HMG-CoAR) and LDL receptor (LDLR) [9]. Conversely, efflux of cholesterol from the cell is achieved with promotion of NPC and ABC transporters. NPC1 and 2 are functionally distinct proteins located in the late endosome that are integral to movement of cholesterol from the endosomal system [86]. ABCA1 and ABCG1 are key regulators of cholesterol transport from cells that mediate the efflux of excess cellular cholesterol and phospholipids from the cell membrane to apolipoprotein A-I (apoA-I) in the plasma, hence decreasing membrane order [87]. Very little is known about the direct role of LXR in cellular GSL metabolism. Deletion of the downstream target gene *npc1* leads to cellular accumulation of gangliosides and other GSLs as well as cholesterol [88]. In addition, sphingolipid metabolism is linked to sterol production via SREBP, the activation of which is also regulated by LXR [89, 90].

LXR in Immunity and Disease

LXR not only responds to oxysterol stimulation to prevent cellular and membrane lipid overload but also modulates a broad spectrum of immune effector responses [91] and regulates T-cell proliferation and immunity [92-95]. This indicates that immune cell functional capacity may be regulated by mechanisms controlling lipid raft composition [85].

Research into the effect of LXRβ on T cell immunology requires further characterization since experiments in murine models have reported both anti-inflammatory [92, 96, 97] and pro-inflammatory [98] results upon activation with synthetic oxysterol agonists such as GW3965 and TO901317. LXR activity inhibits the differentiation and function of Th1, Th2, Th17 and regulatory T cell subsets. The mechanism of action remains to be fully explored but LXR stimulation during T cell activation inhibits proliferation [92]. Furthermore, *in vitro* work on human $CD4^{+}$T cells has indicated a role for LXR in inhibition of chemokine induced cell migration and reduced actin polymerization possibly inhibiting IS formation [94]. In macrophages a convincing inhibitory role for LXR has been revealed via the trans-repression of inducible (i)NOS, IL6 and IL1β and the chemokine monocyte chemoattractant protein-1 (MPC1-), MCP-3 and MMP9 [99, 100].

The role of LXR in the modulation of disease development and progression has been widely studied [101]. The contribution of LXR to atherosclerosis has mainly concentrated on the control of macrophage contribution to plaque development. LXR is required for maximal egress of monocyte-derived cells and lesion regression of plaques in murine models of the disease. Activation of LXR with the synthetic agonist TO901317 improved plasma lipoprotein levels and induced CCR7-dependent lesion regression [102, 103]. In contrast, combined deficiency of the LXR target genes *abca1* and *abcg1* promotes murine foam cell accumulation and plaque progression. These mice also displayed impaired cholesterol efflux, increased inflammatory cytokine and chemokine secretion and a decreased threshold for apoptosis upon free cholesterol or oxidized low density lipoprotein (oxLDL) loading [102, 103].

LXR has also been shown to play a role in autoimmune settings, in particular through regulation of Th17 lymphocyte function. Th17 cells, characterized by the expression of IL-17 and IL-22, are central to the clearance of certain fungal and bacterial pathogens. They have, however, been implicated in the pathogenesis of various autoimmune diseases [104]. Early reports of the inhibition of experimental allergic encephalomyelitis (EAE) development by T0901317-mediated suppression of IL-17, IL-21 and IL-22 suggested a role for LXR in the regulation of Th17 cells [97]. This has since been corroborated by both *in vitro* and *in vivo* work; overexpression of LXR inhibits differentiation of naïve $CD4^{+}T$ cells to the Th17 lineage, whereas and LXR deficiency promotes differentiation. *In vivo* study of the murine EAE model suggested a selective role for LXR-mediated Th17 suppression, as agonist treatment inhibited IL-17 secretion but not IFN-γ in the CNS. The authors suggest LXR mediated inhibition of Th17 differentiation occurs through negative interaction between the lipid regulator SREBP-1 and aryl hydrocarbon receptor (Ahr) [105]. In rheumatoid arthritis, LXR has been shown to be highly upregulated in synovial macrophages, and to promote articular inflammation in part via TLR-driven cytokine release [106]. As described above, our group has recently found perturbed LXR expression and control of lipid metabolism in T cells from patients with SLE. Elevated GSLs in patients with SLE were associated with increased expression of LXRβ an effect recapitulated in healthy $CD4^{+}T$ cells by in *vitro* stimulation with synthetic and endogenous LXR agonists [82].

Whether the effects of LXR stimulation on immune responses in health and disease are mediated via the alteration of plasma membrane lipid rafts remains to be determined.

Lipids Rafts as Therapeutic Targets

Statins reduce cellular cholesterol levels via blocking of the enzyme 3-hydroxy-3-methylglutaryl-coenzyme A (HMG-CoA) reductase. This consequently results in altered membrane cholesterol and disruption of lipid rafts. However, HMG-CoA reductase inhibitors also inhibit lipid-modification of membrane-associated proteins via suppression of protein prenylation. This was shown to underlie the mechanism of action of statins particularly in Th1 response associated animal models [107]. Therefore, statins can alter cell function by affecting both the lipid and protein content of lipid rafts . Statins are known to modulate T cell responses. In T cells isolated from patients with SLE Atorvastatin restored Lck expression and lipid raft-associated signalling [72]. Atorvastatin also inhibits autoreactive B cell activation and delays lupus development in New Zealand black/white F1 mice [108]. In animal models, statins have also been shown to skew the T helper profile towards a more protective Th2 phenotype [109]. Similar to statins, polyunsaturated fatty acids (PUFA) change the structure of lipid rafts via alteration of the raft inner leaflet leading to de-localisation of key signalling molecules, such as Src kinase, form rafts. As a consequence PUFA treatment results in inhibition of T cell activation [110, 111].

The ability of inhibitors of GSL biosynthesis to restore defects in lipid (GSL and cholesterol) homeostasis, inhibit T cell proliferation, cytokine production and differentiation has been identified previously in cell lines and animal models [69, 112]. N-butyldeoxynojirimycin (NB-DNJ) is an N-alkylated imino sugar that functions as an inhibitor of glycosylceramide synthase (GCS). NB-DNJ is employed as a clinical therapy for the treatment of the Gaucher disease, a lysosomal storage disease characterised by an accumulation of glucosylceramide, which manifests in multi-organ dysfunction [113]. Our lab has shown recently that inhibition of GSL biosynthesis *in vitro* using NB-DNJ normalized GSL metabolism, corrected $CD4^{+}T$ cell signaling and functional defects and decreased anti-dsDNA antibody production by autologous B cells in SLE patients. Thus demonstrating a role for defects in metabolism of lipid raft associated GSLs in SLE pathogenesis that could be normalized by targeting GSL biosynthesis [82].

A further example of the therapeutic advantage of GSL inhibitors is Tamoxifen, a major hormonotherapeutic intervention for estrogen receptor positive breast cancer. Tamoxifen functions chiefly as a competitive antagonist to the estrogen receptor, however other targets of the drug such as lipid metabolism have been identified that could account for its biological effects. Tamoxifen was shown to potently reduce levels of glucosylceramide, lactosylceramide and gangliosides [114]. Overexpression of GCS contributes to resistance of cancer cells to chemotherapy, reduction of GCS activity by Tamoxifen was shown to downregulate p-Glycoprotein and reverse multidrug resistance in breast cancer cells, suggesting an important role of lipid metabolism in drug resistance but also suggesting that Tamoxifen could influence lipid raft composition and function [115].

Influenza virus sialidase assists viral infection in the release of virion progeny from infected cells and their movement through the upper respiratory tract. Recent threat of a major human influenza pandemic has prompted the development of novel therapeutic interventions. The most successful of these have been two inhibitors of influenza viral sialidase enzyme, zanamivir and oseltamivir. The inhibitors work by blocking cleavage of α-ketosidically linked residues of glycolipids by viral sialidases, in particular oseltamivir decreases T cell GM1 expression highlighting the importance of lipid rafts in antiviral immunity [35, 116].

Interestingly, a role for the GSL Gb3 in HIV resistance has been recognized [75-77]. Recent work has explored the possibility of blocking HIV infection using a Gb3 analogue [117]. In addition, a novel soluble mimic of the glycolipid, Gb3 inhibits HIV infection [118].

Beyond immune cells, changes in lipid rafts could also be used as a diagnostic tool in neuropathological disorders. Alterations in lipid composition of rafts in disease settings such as Alzheimer's and Parkinson's result in increased lipid raft order, and have been linked to disruption of signalosome activities and potentially neurodegenerative progression. Interestingly, these alterations are significantly perturbed in the first stages of both diseases, and lipid-raft related markers are currently being explored as early diagnostic tools [119].

Conclusion

A growing body of evidence supports a role for membrane lipids in the maintenance of immune cell function and differentiation. Alterations to lipid metabolism driven, for example, by an inflammatory environment, dyslipidemia or infection, could trigger alterations in the maintenance of membrane lipid rafts and their constituents. Taken together the findings discussed above call for a closer investigation to identify the mechanisms by which different lipids can influence immune cell function under normal or pathological conditions. Elucidating such mechanisms will provide the opportunity to devise novel methods of intervention to control aberrant immune cell responses. This raises the exciting possibility that targeting membrane lipids may be an important therapeutic mechanism to control immune responses in a number of diseases.

References

[1] Owen, D. M., Neil, M. A., French, P. M., and Magee, A. I. 2007. Optical techniques for imaging membrane lipid microdomains in living cells. *Semin. Cell Dev. Biol.* 18:591-598.

[2] Zacharias, D. A., Violin, J. D., Newton, A. C., and Tsien, R. Y. 2002. Partitioning of Lipid-Modified Monomeric GFPs into Membrane Microdomains of Live Cells. *Science* 296:913-916.

[3] Munro, S. 2003. Lipid rafts: elusive or illusive? *Cell* 115:377-388.

[4] Shaw, A. S. 2006. Lipid rafts: now you see them, now you don't. *Nat. Immunol.* 7:1139-1142.

[5] Lingwood, D., and Simons, K. 2010. Lipid rafts as a membrane-organizing principle. *Science* 327:46-50.

[6] Varma, R., and Mayor, S. 1998. GPI-anchored proteins are organized in submicron domains at the cell surface. *Nature* 394:798-801.

[7] Schutz, G. J., Kada, G., Pastushenko, V. P., and Schindler, H. 2000. Properties of lipid microdomains in a muscle cell membrane visualized by single molecule microscopy. *EMBO J.* 19:892-901.

[8] Pralle, A., Keller, P., Florin, E. L., Simons, K., and Horber, J. K. 2000. Sphingolipid-cholesterol rafts diffuse as small entities in the plasma membrane of mammalian cells. *J. Cell Biol.* 148:997-1008.

[9] Ikonen, E. 2008. Cellular cholesterol trafficking and compartmentalization. *Nat. Rev. Mol. Cell Biol.* 9:125-138.

[10] Pike, L. J. 2003. Lipid rafts: bringing order to chaos. *J. Lipid Res.* 44:655-667.

[11] Janes, P. W., Ley, S. C., Magee, A. I., and Kabouridis, P. S. 2000. The role of lipid rafts in T cell antigen receptor (TCR) signalling. *Semin. Immunol.* 12:23-34.

[12] Kovarova, M., Tolar, P., Arudchandran, R., Draberova, L., Rivera, J., and Draber, P. 2001. Structure-function analysis of Lyn kinase association with lipid rafts and initiation of early signaling events after Fcepsilon receptor I aggregation. *Mol. Cell Biol.* 21:8318-8328.

[13] Jin, L., Millard, A. C., Wuskell, J. P., Dong, X., Wu, D., Clark, H. A., and Loew, L. M. 2006. Characterization and application of a new optical probe for membrane lipid domains. *Biophys. J.* 90:2563-2575.

[14] Miguel, L., Owen, D. M., Lim, C., Liebig, C., Evans, J., Magee, A. I., and Jury, E. C. 2011. Primary human CD4+ T cells have diverse levels of membrane lipid order that correlate with their function. *J. Immunol.* 186:3505-3516.

[15] Balamuth, F., Leitenberg, D., Unternaehrer, J., Mellman, I., and Bottomly, K. 2001. Distinct patterns of membrane microdomain partitioning in Th1 and th2 cells. *Immunity* 15:729-738.

[16] Owen, D. M., Williamson, D. J., Magenau, A., and Gaus, K. 2012. Sub-resolution lipid domains exist in the plasma membrane and regulate protein diffusion and distribution. *Nat. Commun.* 3:1256.

[17] Ludford-Menting, M. J., Crimeen-Irwin, B., Oliaro, J., Pasam, A., Williamson, D., Pedersen, N., Guillaumot, P., Christansen, D., Manie,

S., Gaus, K., et al. 2011. The Reorientation of T-Cell Polarity and Inhibition of Immunological Synapse Formation by CD46 Involves Its Recruitment to Lipid Rafts. *J. Lipids* 2011:521863.

[18] Tettamanti, G. 2004. Ganglioside/glycosphingolipid turnover: new concepts. *Glycoconj. J.* 20:301-317.

[19] Gault, C. R., Obeid, L. M., and Hannun, Y. A. 2010. An overview of sphingolipid metabolism: from synthesis to breakdown. *Adv. Exp. Med. Biol.* 688:1-23.

[20] Platt, F. M., Boland, B., and van der Spoel, A. C. 2012. The cell biology of disease: lysosomal storage disorders: the cellular impact of lysosomal dysfunction. *J. Cell Biol.* 199:723-734.

[21] Pontier, S. M., and Schweisguth, F. 2012. Glycosphingolipids in signaling and development: from liposomes to model organisms. *Dev. Dyn.* 241:92-106.

[22] Guo, B., Kato, R. M., Garcia-Lloret, M., Wahl, M. I., and Rawlings, D. J. 2000. Engagement of the human pre-B cell receptor generates a lipid raft-dependent calcium signaling complex. *Immunity* 13:243-253.

[23] Ebert, P. J., Baker, J. F., and Punt, J. A. 2000. Immature CD4+CD8+ thymocytes do not polarize lipid rafts in response to TCR-mediated signals. *J. Immunol.* 165:5435-5442.

[24] Simons, K., and Toomre, D. 2000. Lipid rafts and signal transduction. *Nat. Rev. Mol. Cell Biol.* 1:31-39.

[25] Pierce, S. K. 2002. Lipid rafts and B-cell activation. *Nat. Rev. Immunol.* 2:96-105.

[26] Sproul, T. W., Malapati, S., Kim, J., and Pierce, S. K. 2000. Cutting edge: B cell antigen receptor signaling occurs outside lipid rafts in immature B cells. *J. Immunol.* 165:6020-6023.

[27] Tavano, R., Gri, G., Molon, B., Marinari, B., Rudd, C. E., Tuosto, L., and Viola, A. 2004. CD28 and lipid rafts coordinate recruitment of Lck to the immunological synapse of human T lymphocytes. *J. Immunol.* 173:5392-5397.

[28] Viola, A., Schroeder, S., Sakakibara, Y., and Lanzavecchia, A. 1999. T lymphocyte costimulation mediated by reorganization of membrane microdomains. *Science* 283:680-682.

[29] Miceli, M. C., Moran, M., Chung, C. D., Patel, V. P., Low, T., and Zinnanti, W. 2001. Co-stimulation and counter-stimulation: lipid raft clustering controls TCR signaling and functional outcomes. *Semin Immunol* 13:115-128.

[30] Varma, R., Campi, G., Yokosuka, T., Saito, T., and Dustin, M. L. 2006. T cell receptor-proximal signals are sustained in peripheral microclusters and terminated in the central supramolecular activation cluster. *Immunity* 25:117-127.
[31] Ballek, O., Brouckova, A., Manning, J., and Filipp, D. 2012. A specific type of membrane microdomains is involved in the maintenance and translocation of kinase active Lck to lipid rafts. *Immunol. Lett.* 142:64-74.
[32] Zhang, M., Moran, M., Round, J., Low, T. A., Patel, V. P., Tomassian, T., Hernandez, J. D., and Miceli, M. C. 2005. CD45 signals outside of lipid rafts to promote ERK activation, synaptic raft clustering, and IL-2 production. *J. Immunol.* 174:1479-1490.
[33] van der Goot, F. G., Tran van Nhieu, G., Allaoui, A., Sansonetti, P., and Lafont, F. 2004. Rafts can trigger contact-mediated secretion of bacterial effectors via a lipid-based mechanism. *J Biol Chem* 279:47792-47798.
[34] Blery, M., Tze, L., Miosge, L. A., Jun, J. E., and Goodnow, C. C. 2006. Essential role of membrane cholesterol in accelerated BCR internalization and uncoupling from NF-kappa B in B cell clonal anergy. *J. Exp. Med.* 203:1773-1783.
[35] Moore, M. L., Chi, M. H., Zhou, W., Goleniewska, K., O'Neal, J. F., Higginbotham, J. N., and Peebles, R. S., Jr. 2007. Cutting Edge: Oseltamivir decreases T cell GM1 expression and inhibits clearance of respiratory syncytial virus: potential role of endogenous sialidase in antiviral immunity. *J. Immunol.* 178:2651-2654.
[36] Roy, K., Ghosh, M., Pal, T. K., Chakrabarti, S., and Roy, S. 2013. Cholesterol lowering drug may influence cellular immune response by altering MHC II function. *J. Lipid. Res.* 54:3106-3115.
[37] Knorr, R., Karacsonyi, C., and Lindner, R. 2009. Endocytosis of MHC molecules by distinct membrane rafts. *J. Cell Sci.* 122:1584-1594.
[38] Gombos, I., Kiss, E., Detre, C., Laszlo, G., and Matko, J. 2006. Cholesterol and sphingolipids as lipid organizers of the immune cells' plasma membrane: their impact on the functions of MHC molecules, effector T-lymphocytes and T-cell death. *Immunol. Lett.* 104:59-69.
[39] Berzins, S. P., Smyth, M. J., and Baxter, A. G. 2011. Presumed guilty: natural killer T cell defects and human disease. *Nat. Rev. Immunol.* 11:131-142.
[40] Im, J. S., Arora, P., Bricard, G., Molano, A., Venkataswamy, M. M., Baine, I., Jerud, E. S., Goldberg, M. F., Baena, A., Yu, K. O., et al.

2009. Kinetics and cellular site of glycolipid loading control the outcome of natural killer T cell activation. *Immunity* 30:888-898.

[41] Vainio, S., Jansen, M., Koivusalo, M., Rog, T., Karttunen, M., Vattulainen, I., and Ikonen, E. 2006. Significance of sterol structural specificity. Desmosterol cannot replace cholesterol in lipid rafts. *J. Biol. Chem.* 281:348-355.

[42] Megha, Bakht, O., and London, E. 2006. Cholesterol Precursors Stabilize Ordinary and Ceramide-rich Ordered Lipid Domains (Lipid Rafts) to Different Degrees: Implications For The Bloch Hypothesis And Sterol Biosynthesis Disorders. *Journal of Biological Chemistry* 281:21903-21913.

[43] Miguel, L., Owen, D. M., Lim, C., Liebig, C., Evans, J., Magee, A. I., and Jury, E. C. 2011. Primary Human CD4+ T Cells Have Diverse Levels of Membrane Lipid Order That Correlate with Their Function. *The Journal of Immunology* 186:3505-3516.

[44] Larbi, A., Douziech, N., Khalil, A., Dupuis, G., Gherairi, S., Guerard, K. P., and Fulop, T., Jr. 2004. Effects of methyl-beta-cyclodextrin on T lymphocytes lipid rafts with aging. *Exp. Gerontol.* 39:551-558.

[45] Nguyen, D. H., Espinoza, J. C., and Taub, D. D. 2004. Cellular cholesterol enrichment impairs T cell activation and chemotaxis. *Mech. Ageing Dev.* 125:641-650.

[46] Kiguchi, K., Henning-Chubb, C. B., and Huberman, E. 1990. Glycosphingolipid patterns of peripheral blood lymphocytes, monocytes, and granulocytes are cell specific. *J. Biochem.* 107:8-14.

[47] Schwarting, G. A. 1980. Quantitative analysis of neutral glycosphingolipids from human lymphocyte subpopulations. *Biochem. J.* 189:407-412.

[48] Sorice, M., Longo, A., Garofalo, T., Mattei, V., Misasi, R., and Pavan, A. 2004. Role of GM3-enriched microdomains in signal transduction regulation in T lymphocytes. *Glycoconj. J.* 20:63-70.

[49] Garofalo, T., Lenti, L., Longo, A., Misasi, R., Mattei, V., Pontieri, G. M., Pavan, A., and Sorice, M. 2002. Association of GM3 with Zap-70 induced by T cell activation in plasma membrane microdomains: GM3 as a marker of microdomains in human lymphocytes. *J. Biol. Chem.* 277:11233-11238.

[50] Garofalo, T., Sorice, M., Misasi, R., Cinque, B., Mattei, V., Pontieri, G. M., Cifone, M. G., and Pavan, A. 2002. Ganglioside GM3 activates ERKs in human lymphocytic cells. *J. Lipid Res.* 43:971-978.

[51] Morales, A., Colell, A., Mari, M., Garcia-Ruiz, C., and Fernandez-Checa, J. C. 2004. Glycosphingolipids and mitochondria: role in apoptosis and disease. *Glycoconj. J.* 20:579-588.

[52] Degroote, S., Wolthoorn, J., and van Meer, G. 2004. The cell biology of glycosphingolipids. *Semin Cell Dev. Biol.* 15:375-387.

[53] Janes, P. W., Ley, S. C., and Magee, A. I. 1999. Aggregation of lipid rafts accompanies signaling via the T cell antigen receptor. *J. Cell Biol.* 147:447-461.

[54] Nagafuku, M., Okuyama, K., Onimaru, Y., Suzuki, A., Odagiri, Y., Yamashita, T., Iwasaki, K., Fujiwara, M., Takayanagi, M., Ohno, I., et al. 2012. CD4 and CD8 T cells require different membrane gangliosides for activation. *Proc. Natl. Acad. Sci. U. S. A.* 109:E336-342.

[55] Shu, L., Lee, L., and Shayman, J. A. 2002. Regulation of phospholipase C-gamma activity by glycosphingolipids. *J. Biol. Chem.* 277:18447-18453.

[56] Gomez-Mouton, C., Abad, J. L., Mira, E., Lacalle, R. A., Gallardo, E., Jimenez-Baranda, S., Illa, I., Bernad, A., Manes, S., and Martinez, A. C. 2001. Segregation of leading-edge and uropod components into specific lipid rafts during T cell polarization. *Proc. Natl. Acad. Sci. U. S. A.* 98:9642-9647.

[57] Schade, A. E., and Levine, A. D. 2002. Lipid raft heterogeneity in human peripheral blood T lymphoblasts: a mechanism for regulating the initiation of TCR signal transduction. *J. Immunol.* 168:2233-2239.

[58] Barbat, C., Trucy, M., Sorice, M., Garofalo, T., Manganelli, V., Fischer, A., and Mazerolles, F. 2007. p56lck, LFA-1 and PI3K but not SHP-2 interact with GM1- or GM3-enriched microdomains in a CD4-p56lck association-dependent manner. *Biochem. J.* 402:471-481.

[59] Sorice, M., Garofalo, T., Misasi, R., Longo, A., Mikulak, J., Dolo, V., Pontieri, G.M., and Pavan, A. 2000. Association between GM3 and CD4-Ick complex in human peripheral blood lymphocytes. *Glycoconj. J.* 17:247-252.

[60] Inokuchi, J. I., Nagafuku, M., Ohno, I., and Suzuki, A. 2012. Heterogeneity of gangliosides among T cell subsets. *Cell Mol. Life Sci.*

[61] Rudd, C. E. 2008. The reverse stop-signal model for CTLA4 function. *Nat. Rev. Immunol.* 8:153-160.

[62] Martin, M., Schneider, H., Azouz, A., and Rudd, C. E. 2001. Cytotoxic T lymphocyte antigen 4 and CD28 modulate cell surface raft expression in their regulation of T cell function. *J. Exp. Med.* 194:1675-1681.

[63] Chikuma, S., Imboden, J. B., and Bluestone, J. A. 2003. Negative regulation of T cell receptor-lipid raft interaction by cytotoxic T lymphocyte-associated antigen 4. *J. Exp. Med.* 197:129-135.

[64] Jury, E. C., Flores-Borja, F., and Kabouridis, P. S. 2007. Lipid rafts in T cell signalling and disease. *Semin. Cell Dev. Biol.* 18:608-615.

[65] Allen, J. A., Halverson-Tamboli, R. A., and Rasenick, M. M. 2007. Lipid raft microdomains and neurotransmitter signalling. *Nat. Rev. Neurosci.* 8:128-140.

[66] Muppidi, J. R., and Siegel, R. M. 2004. Ligand-independent redistribution of Fas (CD95) into lipid rafts mediates clonotypic T cell death. *Nat. Immunol.* 5:182-189.

[67] Rentero, C., Zech, T., Quinn, C. M., Engelhardt, K., Williamson, D., Grewal, T., Jessup, W., Harder, T., and Gaus, K. 2008. Functional implications of plasma membrane condensation for T cell activation. *PLoS One* 3:e2262.

[68] Massey, J. B., and Pownall, H. J. 2005. The polar nature of 7-ketocholesterol determines its location within membrane domains and the kinetics of membrane microsolubilization by apolipoprotein A-I. *Biochemistry* 44:10423-10433.

[69] Zhu, Y., Gumlaw, N., Karman, J., Zhao, H., Zhang, J., Jiang, J. L., Maniatis, P., Edling, A., Chuang, W. L., Siegel, C., et al. 2011. Lowering glycosphingolipid levels in CD4+ T cells attenuates T cell receptor signaling, cytokine production, and differentiation to the Th17 lineage. *J. Biol. Chem.* 286:14787-14794.

[70] Flores-Borja, F., Kabouridis, P. S., Jury, E. C., Isenberg, D. A., and Mageed, R. A. 2007. Altered lipid raft-associated proximal signaling and translocation of CD45 tyrosine phosphatase in B lymphocytes from patients with systemic lupus erythematosus. *Arthritis Rheum.* 56:291-302.

[71] Jury, E. C., Kabouridis, P. S., Flores-Borja, F., Mageed, R. A., and Isenberg, D. A. 2004. Altered lipid raft-associated signaling and ganglioside expression in T lymphocytes from patients with systemic lupus erythematosus. *J. Clin. Invest.* 113:1176-1187.

[72] Jury, E. C., Isenberg, D. A., Mauri, C., and Ehrenstein, M. R. 2006. Atorvastatin restores Lck expression and lipid raft-associated signaling in T cells from patients with systemic lupus erythematosus. *J. Immunol.* 177:7416-7422.

[73] Scheiffele, P., Rietveld, A., Wilk, T., and Simons, K. 1999. Influenza viruses select ordered lipid domains during budding from the plasma membrane. *J. Biol. Chem.* 274:2038-2044.
[74] Zhang, J., Pekosz, A., and Lamb, R. A. 2000. Influenza virus assembly and lipid raft microdomains: a role for the cytoplasmic tails of the spike glycoproteins. *J. Virol.* 74:4634-4644.
[75] Dykstra, M., Cherukuri, A., Sohn, H. W., Tzeng, S. J., and Pierce, S. K. 2003. Location is everything: lipid rafts and immune cell signaling. *Annu. Rev. Immunol.* 21:457-481.
[76] Ono, A., and Freed, E. O. 2001. Plasma membrane rafts play a critical role in HIV-1 assembly and release. *Proc. Natl. Acad. Sci. U. S. A.* 98:13925-13930.
[77] Hug, P., Lin, H. M., Korte, T., Xiao, X., Dimitrov, D. S., Wang, J. M., Puri, A., and Blumenthal, R. 2000. Glycosphingolipids promote entry of a broad range of human immunodeficiency virus type 1 isolates into cell lines expressing CD4, CXCR4, and/or CCR5. *J. Virol.* 74:6377-6385.
[78] Alfsen, A., Iniguez, P., Bouguyon, E., and Bomsel, M. 2001. Secretory IgA specific for a conserved epitope on gp41 envelope glycoprotein inhibits epithelial transcytosis of HIV-1. *J. Immunol.* 166:6257-6265.
[79] Nan, X., Carubelli, I., and Stamatos, N. M. 2007. Sialidase expression in activated human T lymphocytes influences production of IFN-gamma. *J. Leukoc. Biol.* 81:284-296.
[80] Barrias, E. S., Dutra, J. M., De Souza, W., and Carvalho, T. M. 2007. Participation of macrophage membrane rafts in Trypanosoma cruzi invasion process. *Biochem. Biophys. Res. Commun.* 363:828-834.
[81] Jury, E. C., Flores-Borja, F., Kalsi, H. S., Lazarus, M., Isenberg, D. A., Mauri, C., and Ehrenstein, M. R. 2010. Abnormal CTLA-4 function in T cells from patients with systemic lupus erythematosus. *Eur. J. Immunol.* 40:569-578.
[82] McDonald, G., Deepak, S., Miguel, L., Hall, C. J., Isenberg, D. A., Magee, A. I., Butters, T., Jury, E. C. (2013) Normalizing glycosphingolipid turnover restores function in SLE CD4+ T cells. Journal of Clinical Investigation. In press.
[83] Kabouridis, P. S., and Jury, E. C. 2008. Lipid rafts and T-lymphocyte function: implications for autoimmunity. *FEBS Lett.* 582:3711-3718.
[84] Lemaire-Ewing, S., Lagrost, L., and Neel, D. 2012. Lipid rafts: a signalling platform linking lipoprotein metabolism to atherogenesis. *Atherosclerosis* 221:303-310.

[85] Zelcer, N., and Tontonoz, P. 2006. Liver X receptors as integrators of metabolic and inflammatory signaling. *J. Clin. Invest.* 116:607-614.

[86] Subramanian, K., and Balch, W. E. 2008. NPC1/NPC2 function as a tag team duo to mobilize cholesterol. *Proc. Natl. Acad. Sci. U. S. A.* 105:15223-15224.

[87] Rees, D. C., Johnson, E., and Lewinson, O. 2009. ABC transporters: the power to change. *Nat. Rev. Mol. Cell Biol.* 10:218-227.

[88] Vanier, M.T. 1999. Lipid changes in Niemann-Pick disease type C brain: personal experience and review of the literature. *Neurochem. Res.* 24:481-489.

[89] Yoshikawa, T., Shimano, H., Amemiya-Kudo, M., Yahagi, N., Hasty, A. H., Matsuzaka, T., Okazaki, H., Tamura, Y., Iizuka, Y., Ohashi, K., et al. 2001. Identification of liver X receptor-retinoid X receptor as an activator of the sterol regulatory element-binding protein 1c gene promoter. *Mol. Cell Biol.* 21:2991-3000.

[90] Breslow, D. K., and Weissman, J. S. 2010. Membranes in balance: mechanisms of sphingolipid homeostasis. *Mol. Cell* 40:267-279.

[91] Steffensen, K. R., Jakobsson, T., and Gustafsson, J. A. 2013. Targeting liver X receptors in inflammation. *Expert Opin. Ther. Targets* 17:977-990.

[92] Bensinger, S. J., Bradley, M. N., Joseph, S. B., Zelcer, N., Janssen, E. M., Hausner, M. A., Shih, R., Parks, J. S., Edwards, P. A., Jamieson, B. D., et al. 2008. LXR signaling couples sterol metabolism to proliferation in the acquired immune response. *Cell* 134:97-111.

[93] Walcher, D., Kummel, A., Kehrle, B., Bach, H., Grub, M., Durst, R., Hombach, V., and Marx, N. 2006. LXR activation reduces proinflammatory cytokine expression in human CD4-positive lymphocytes. *Arterioscler. Thromb. Vasc. Biol.* 26:1022-1028.

[94] Walcher, D., Vasic, D., Heinz, P., Bach, H., Durst, R., Hausauer, A., Hombach, V., and Marx, N. 2010. LXR activation inhibits chemokine-induced CD4-positive lymphocyte migration. *Basic Res. Cardiol.* 105:487-494.

[95] Geyeregger, R., Shehata, M., Zeyda, M., Kiefer, F. W., Stuhlmeier, K. M., Porpaczy, E., Zlabinger, G. J., Jager, U., and Stulnig, T. M. 2009. Liver X receptors interfere with cytokine-induced proliferation and cell survival in normal and leukemic lymphocytes. *J. Leukoc. Biol.* 86:1039-1048.

[96] Solt, L. A., Kamenecka, T. M., and Burris, T. P. 2012. LXR-mediated inhibition of CD4+ T helper cells. *PLoS One* 7:e46615.

[97] Xu, J., Wagoner, G., Douglas, J. C., and Drew, P. D. 2009. Liver X receptor agonist regulation of Th17 lymphocyte function in autoimmunity. *J. Leukoc. Biol.* 86:401-409.
[98] Asquith, D. L., Miller, A. M., Hueber, A. J., McKinnon, H. J., Sattar, N., Graham, G. J., and McInnes, I. B. 2009. Liver X receptor agonism promotes articular inflammation in murine collagen-induced arthritis. *Arthritis Rheum.* 60:2655-2665.
[99] Joseph, S. B., Castrillo, A., Laffitte, B. A., Mangelsdorf, D. J., and Tontonoz, P. 2003. Reciprocal regulation of inflammation and lipid metabolism by liver X receptors. *Nat. Med.* 9:213-219.
[100] Castrillo, A., Joseph, S. B., Marathe, C., Mangelsdorf, D. J., and Tontonoz, P. 2003. Liver X receptor-dependent repression of matrix metalloproteinase-9 expression in macrophages. *J. Biol. Chem.* 278:10443-10449.
[101] Poli, G., Biasi, F., and Leonarduzzi, G. 2013. Oxysterols in the pathogenesis of major chronic diseases. *Redox. Biol.* 1:125-130.
[102] van der Stoep, M., Li, Z., Calpe-Berdiel, L., van der Sluis, R. J., Saleh, P., McKinnon, H. J., Smit, M. J., Korporaal, S. J., Van Berkel, T. J., Van Eck, M., et al. 2013. Elimination of macrophages drives LXR-induced regression both in initial and advanced stages of atherosclerotic lesion development. *Biochem. Pharmacol.*
[103] Feig, J. E., Pineda-Torra, I., Sanson, M., Bradley, M. N., Vengrenyuk, Y., Bogunovic, D., Gautier, E. L., Rubinstein, D., Hong, C., Liu, J., et al. 2010. LXR promotes the maximal egress of monocyte-derived cells from mouse aortic plaques during atherosclerosis regression. *J. Clin. Invest.* 120:4415-4424.
[104] Korn, T., Bettelli, E., Oukka, M., and Kuchroo, V. K. 2009. IL-17 and Th17 Cells. *Annu. Rev. Immunol.* 27:485-517.
[105] Cui, G., Qin, X., Wu, L., Zhang, Y., Sheng, X., Yu, Q., Sheng, H., Xi, B., Zhang, J. Z., and Zang, Y. Q. 2011. Liver X receptor (LXR) mediates negative regulation of mouse and human Th17 differentiation. *J. Clin. Invest.* 121:658-670.
[106] Asquith, D. L., Ballantine, L. E., Nijjar, J. S., Makdasy, M. K., Patel, S., Wright, P. B., Reilly, J. H., Kerr, S., Kurowska-Stolarska, M., Gracie, J. A., et al. 2013. The liver X receptor pathway is highly upregulated in rheumatoid arthritis synovial macrophages and potentiates TLR-driven cytokine release. *Ann. Rheum. Dis.*
[107] Dunn, S. E., Youssef, S., Goldstein, M. J., Prod'homme, T., Weber, M. S., Zamvil, S. S., and Steinman, L. 2006. Isoprenoids determine

Th1/Th2 fate in pathogenic T cells, providing a mechanism of modulation of autoimmunity by atorvastatin. *J. Exp. Med.* 203:401-412.

[108] Lawman, S., Mauri, C., Jury, E. C., Cook, H. T., and Ehrenstein, M. R. 2004. Atorvastatin inhibits autoreactive B cell activation and delays lupus development in New Zealand black/white F1 mice. *J. Immunol.* 173:7641-7646.

[109] Youssef, S., Stuve, O., Patarroyo, J. C., Ruiz, P. J., Radosevich, J. L., Hur, E. M., Bravo, M., Mitchell, D. J., Sobel, R. A., Steinman, L., et al. 2002. The HMG-CoA reductase inhibitor, atorvastatin, promotes a Th2 bias and reverses paralysis in central nervous system autoimmune disease. *Nature* 420:78-84.

[110] Stulnig, T. M., Berger, M., Sigmund, T., Raederstorff, D., Stockinger, H., and Waldhausl, W. 1998. Polyunsaturated fatty acids inhibit T cell signal transduction by modification of detergent-insoluble membrane domains. *J. Cell Biol.* 143:637-644.

[111] Webb, Y., Hermida-Matsumoto, L., and Resh, M. D. 2000. Inhibition of protein palmitoylation, raft localization, and T cell signaling by 2-bromopalmitate and polyunsaturated fatty acids. *J. Biol. Chem.* 275:261-270.

[112] Glaros, E. N., Kim, W. S., Quinn, C. M., Wong, J., Gelissen, I., Jessup, W., and Garner, B. 2005. Glycosphingolipid accumulation inhibits cholesterol efflux via the ABCA1/apolipoprotein A-I pathway: 1-phenyl-2-decanoylamino-3-morpholino-1-propanol is a novel cholesterol efflux accelerator. *J. Biol. Chem.* 280:24515-24523.

[113] Ficicioglu, C. 2008. Review of miglustat for clinical management in Gaucher disease type 1. *Ther. Clin. Risk Manag.* 4:425-431.

[114] Cabot, M. C., Giuliano, A. E., Volner, A., and Han, T. Y. 1996. Tamoxifen retards glycosphingolipid metabolism in human cancer cells. *FEBS Lett.* 394:129-131.

[115] Gouaze, V., Liu, Y. Y., Prickett, C. S., Yu, J. Y., Giuliano, A. E., and Cabot, M. C. 2005. Glucosylceramide synthase blockade down-regulates P-glycoprotein and resensitizes multidrug-resistant breast cancer cells to anticancer drugs. *Cancer Res.* 65:3861-3867.

[116] von Itzstein, M. 2007. The war against influenza: discovery and development of sialidase inhibitors. *Nat. Rev. Drug Discov.* 6:967-974.

[117] Lund, N., Branch, D. R., Mylvaganam, M., Chark, D., Ma, X. Z., Sakac, D., Binnington, B., Fantini, J., Puri, A., Blumenthal, R., et al. 2006. A novel soluble mimic of the glycolipid, globotriaosyl ceramide inhibits HIV infection. *AIDS* 20:333-343.

[118] Lingwood, C. A., and Branch, D. R. 2011. The role of glycosphingolipids in HIV/AIDS. *Discov. Med.* 11:303-313.
[119] Marin, R., Rojo, J. A., Fabelo, N., Fernandez, C. E., and Diaz, M. 2013. Lipid raft disarrangement as a result of neuropathological progresses: a novel strategy for early diagnosis? *Neuroscience* 245:26-39.
[120] Simons, K., and Ehehalt, R. 2002. Cholesterol, lipid rafts, and disease. *J. Clin. Invest.* 110:597-603.

In: Lipid Rafts
Editor: Dan Sillence

ISBN: 978-1-62948-961-2

Chapter VI

Lipid Rafts and Signal Transduction in Mast Cells

Vivian Marino Mazucato*[*], *Adriana Maria Mariano Silveira e Souza*[†], *Maria Célia Jamur*[‡] and *Constance Oliver*[§]
Department of Cell and Molecular Biology and Pathogenic Bioagents, Faculdade de Medicina de Ribeirão Preto, University of São Paulo, Ribeirão Preto, SP, Brazil

Abstract

In mast cells, the first signaling process convincingly shown to involve lipid rafts was immunoglobulin E (IgE) signaling during an allergic response. The cross-linking of IgE bound to the high affinity IgE receptor (FcεRI), mediated by oligomeric antigen, is one of the key pathways in mast cell activation. Activation of FcεRI stimulates the release of numerous chemical mediators involved in allergic reactions.

[*] Email: vivianmazucato@hotmail.com (VMM).
[†] Email: drisilvesouza@gmail.com (AMMSS).
[‡] Email: mjamur@fmrp.usp.br (MCJ).
[§] Address correspondence to: Dr. Constance Oliver, Department of Cell and Molecular Biology and Pathogenic Bioagents, School of Medicine of Ribeirão Preto - University of São Paulo. Av. Bandeirantes 3900, 14049-900 - Ribeirão Preto, SP, Brazil. Phone: 55-16-3602-3217, FAX: 55-16-3633-1786, e-mail: coliver@fmrp.usp.br.

Mast cells, therefore, are a major source of cytokines and chemokines, similar to other immunoregulatory cells. The cross-linking of FcεRI, activates a transmembrane signaling process and leads to its translocation to specialized membrane micro-domains, known as lipid rafts. Together with FcεRI, other proteins, which are important for the signaling cascade in mast cells, are recruited to these micro-domains. Besides this recruitment, some transmembrane proteins, proteins with lipidic modification, high levels of cholesterol, and gangliosides are also present and support the formation of the signaling platform. Many cellular functions have been connected with lipid rafts including endocytosis, cholesterol transport, as well as signal transduction. Gangliosides are a characteristic component of lipid rafts and have been shown to act in cell-cell recognition, cell-substrate interaction, as receptors for virus and bacterial toxins, and also to have a role in proliferation, differentiation, adhesion, immune response, signal transduction. This chapter provides an overview of the importance of the lipid rafts in signal transduction in mast cells via FcεRI and the importance of lipid rafts in modulating biological processes induced by signaling in these cells.

Introduction

The fluid-mosaic model of cell membranes [1] postulated a uniform double layer of lipids with proteins randomly suspended in the bilayer. This lipid-protein bilayer delimits cells and organelles and acts as a barrier to the passage of polar molecules and ions [2-5].The model predicted free rotational and lateral diffusion of proteins and lipids within the plane of the membrane resulting in their random distribution. A number of lines of evidence indicating a non-random distribution of complex lipids and proteins in cell membranes was presented in seventies and eighties suggested that the fluid-mosaic model was probably inaccurate. Lee and collaborators [6], studying the effects of temperature on membrane behavior, proposed the presence of clusters of lipids in membranes. Wunderlich and collaborators [7] suggested that these clusters might be quasicrystalline regions surrounded by more freely dispersed liquid crystalline lipid molecules. Karnovsky and collaborators [8] formalized the concept of the organization of the lipid components of membranes into domains. However, it was not until the 1990's that membrane fractionation techniques improved sufficiently to provide the initial information on membrane micro-domains, and the concept of lipid rafts was proposed [9-14].

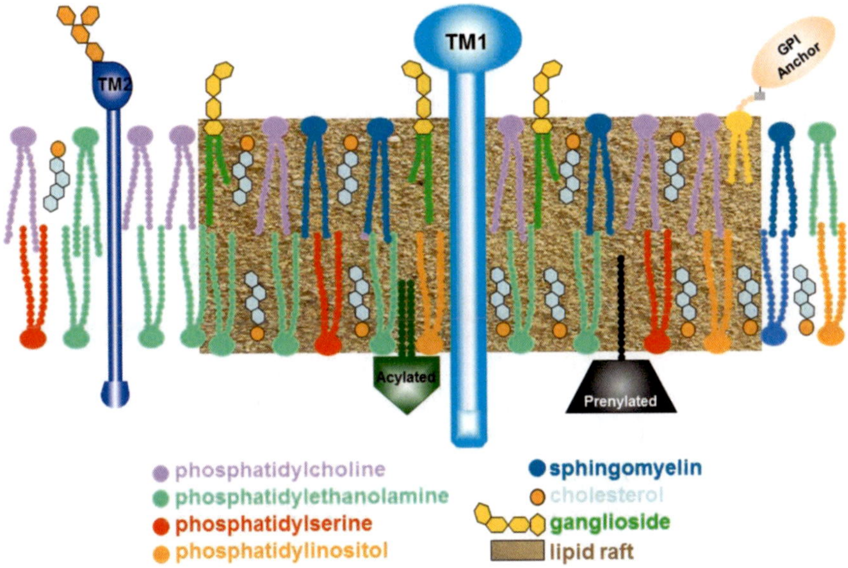

Figure 1. Schematic illustration of a lipid raft. Lipid rafts contain high levels of cholesterol, sphingolipids and gangliosides. GPI anchored proteins along with sphingomyelin and gangliosides are present in the outer leaflet. Acylated and prenylated proteins, phosphatidlyserine and phosphatidylethanolamine are located in the inner leaflet. Cholesterol is present in both leaflets and functions as a space filling molecule under the sphingolipid head groups. Transmembrane proteins may be associated with the lipid raft (TM1) or outside of the lipid raft (TM2).

The lipid rafts (Figure 1), present in all eukaryotic cells, are not a single stable structure in the plasma membrane. They are a dynamic and heterogeneous set of subdomains that differ in protein and lipid composition. Lipid rafts contain high levels of cholesterol (cholesterol levels in rafts are generally double those found in the plasma membranes from which they were derived), sphingolipids (such as sphingomyelin) and gangliosides [15-18]. Also, they are transitory in nature and their composition can change with the physiological state of the cell. The lipid rafts exhibit variability in their shape, size, lifetime and stability [19-21]. Lipid rafts are currently defined as dynamic ordered meso-scale (5-200 nm) assemblies of proteins and lipids of the plasma membrane and other intracellular membranes, such as Golgi membranes, that associate and dissociate on a sub-second timescale [5, 22-25].

A large number of studies suggest that lipid rafts selectively concentrate glycosylphosphatidylinositol (GPI)-anchored proteins on their outer side and proteins with lipid modifications (palmitoyl or myristoyl groups) and cholesterol-binding proteins on the cytoplasmic side [26-30]. On the other hand, proteins with transmembrane segments have been shown to be targeted to lipid rafts by amino acid sequences in their extracellular [31], transmembrane [32], or intracellular domains [33]. However, little progress has been made in determining the nature of these protein-based raft targeting sequences, therefore it is difficult to predict, on the basis of sequence, whether a protein is likely to be directed to rafts [5]. Lipid raft composition, with a preponderance of longer saturated hydrocarbon chains that potentiate interdigitation between leaflets [34] and favors interaction with cholesterol [24], allows cholesterol to be tightly intercalated within the raft. The lipid rafts arc highly organized and probably exist in a liquid-ordered (l_o) phase, different from the rest of the plasma membrane which consists mainly of phospholipids (with unsaturated tails) in a liquid-disordered (l_d) phase [35, 36]. The extent of packing depends on the degree of saturation of the lipid. The cis double bond present on unsaturated lipids introduces a rigid bend in the hydrocarbon tail which interferes with the tight packing and results in less stable aggregates [37]. These membrane micro-domains are characterized by a high melting temperature and resistance to solubilization in non-ionic detergents, such as Triton X-100, at low temperature [38]. They are dynamic in that they can coalesce or fragment into smaller units [30] and both proteins and lipids can move in and out of raft domains with different partitioning kinetics [29]. Lipid rafts can also form stabile platforms that are important in signaling, viral infection and membrane trafficking [23, 39, 40]. The high concentration of two types of the lipids, sphingolipids and cholesterol, confer stability on the lipid rafts formed in the plasma membrane [41]. Despite a body of evidence supporting the existence of raft domains, the raft concept is still being debated [42] and their true nature and even their existence in live cells has been a subject of controversy [5, 43-45]. These controversies concerning the existence of lipid rafts may be due to restrictions in the methods available to isolate and to characterize lipid rafts [46]. Lipid rafts have not been directly observed *in vivo* without substantial nonphysiological perturbation, such as long-term incubation in cold or extensive cross-linking of raft markers [20, 47]. Furthermore, the mechanisms that govern the associations among sphingolipid, cholesterol and specific membrane proteins in live cell membranes remain unclear [48]. The majority of the studies involving lipid rafts begin with detergent solubilization of whole cells followed by sucrose density gradient centrifugation and the recovery of detergent-resistant membranes from the light fractions of the gradient [49, 50]. However, the analysis of density gradient centrifugation experiments remains controversial because there is an indication that detergents may force associations between components that are not associated in intact cells [51]. Fractionation results are also known to be severely altered by varying a number of factors such as the concentration of Triton X-100 [52, 53], the use of different detergents [54, 55] or the omission of detergents in general [56-58]. However, biochemical isolation of subcellular components remains an essential part of structure-function analysis in cells, even while recognizing that these procedures perturb cellular structure relative to those seen in live cells. Isolation of lipid rafts by detergent fractionation or other methods suffers the same pitfalls, but provides the means necessary for molecular characterization that can be further refined in live cell experiments [59]. Recently, compelling data has been presented that self-organization of lipids and proteins can induce a sub-compartmentalization that organizes the bioactivity of cell membranes [24, 25]. Lately, the lipid-based phase separation into liquid-ordered-like and liquid-disordered-like has been seen in giant plasma membrane vesicles (GPMVs) obtained by chemically induced blebbing in cultured cells [60, 61] or by using cell swelling to generate plasma membrane spheres (PMS) [62]. Johnson and collaborators, [63] using GPMVs showed that peripheral protein binding may be a regulator for lateral heterogeneity *in vivo.* These new approaches are very promising, allowing studies of the lipid domains in the absence of detergents. Advances in imaging and studies with improved integrated methodologies, such as flotation of detergent-resistant membranes, antibody-induced patching and immunofluorescence microscopy, immunoelectron microscopy, chemical crosslinking, single fluorophore tracking microscopy, photonic force microscopy, spectrofluorimetry, mass spectrometry, and fluorescence resonance energy transfer (FRET) are now providing insights into the existence and behavior of lipid rafts [3, 39, 47, 64-70]. Proteomic analyses have also provided confirmation of the raft localization of many proteins previously shown to partition into lipid rafts using other methods. These studies have also identified novel proteins in rafts and led to insights into the physiological regulation of rafts [5]. Lipidomic studies of lipid rafts have been done using rafts prepared by both detergent-free and detergent-containing protocols, when these results are compared, significant differences in lipid composition have been identified [15]. Thus, attention is required when the results of individual raft lipidomics studies are interpreted [5].

Studies of lipid rafts have also been complicated by imprecise nomenclature [39]. For example lipid rafts are also known as detergent-resistant membranes (DREMs) or glycosphingolipids-enriched membranes

(GEMs) [2, 14, 71], and caveolae were synonymous with lipid rafts for many years. In 1998, Harder *et al.*, using a cell system lacking caveolae, demonstrated that raft and non-raft markers segregated in the same cholesterol-dependent way in the absence of caveolae. These results showed that clustered raft markers segregate away from non-raft proteins in a cholesterol dependent, but caveolin independent manner [70]. Today caveolae are considered a subset of lipid rafts [3, 72]. Lipid rafts were so named because it was originally thought that they represented pre-existing domains in membranes into which different proteins partitioned [5].

Functional roles for lipid rafts and lipid heterogeneity in cellular processes have received increasing attention during the past decade [73]. These membrane micro-domains have been implicated in a variety of cellular processes, including pro- and anti-apoptotic events, host-pathogen interactions, endocytic events, such as viral entry, vesicle trafficking to or from the plasma membrane, signal transduction, cell adhesion, play roles in the initiation of many pharmacological agent-induced signaling pathways and toxicological effects [74-76]. However, the most important role of rafts at the cell surface may be their function in cell signaling by virtue of their affinity for signaling proteins, including transmembrane receptors, GPI anchored proteins, G proteins, RhoA, linker for activation of T cells (LAT, also used as a lipid raft marker) and Src kinases (Rous asian sarcoma tyrosine kinase) [42, 49]. The number of proteins reported to be regulated by specific lipid interaction is steadily increasing, but the precise structural mechanisms behind specific binding and receptor regulation in membranes remain uncharacterized [70]. Biochemical and genetic data have lent credence to the notion that rafts function as a specialized signaling platform in cell membranes [77-83]. Most likely, the function of the rafts is aided by stimulation-induced association and recruitment of various molecules with raft affinity, as well as varying degrees of raft engagement with the cytoskeleton [30, 71, 84]. Lipid rafts are also thought to be important sites for protein-tyrosine-kinase-mediated protein-protein interactions that are involved in the initiation of receptor signaling pathways [3, 71, 85]. It is well known that, in the case of tyrosine kinase receptors, adaptors, scaffolding proteins and enzymes are recruited to the cytoplasmic side of the plasma membrane as a result of ligand binding to form a signaling complex [86]. If receptor activation takes place in an ordered lipid raft, the signaling complex is protected from other proteins, such as membrane phosphatases, localized in the disordered region of the plasma membrane, that otherwise could affect the signaling process [44, 66, 87]. Lipid rafts serve as a platform for diverse signaling pathways, such as those mediated by growth factors, morphogens, integrins [3] and antigen receptors on immune cells [40, 88-91].

This chapter presents evidence that lipid rafts play a pivotal role in signal transduction in mast cells and that the organization of various molecules in lipid rafts could modulate many biological processes in these cells.

Mast Cells

Mast cells were first recognized by Paul Ehrlich [92]. His microscopic observations revealed that mast cells had a high content of cytoplasmic granules, which stain metachromatically with basic aniline dyes such as toluidine blue (Figure 2).

Like blood cells, mast cells are derived from pluripotent bone marrow hematopoietic stem cells but, unlike other blood cells, which circulate in the blood, mature mast cells are not detected in circulation and reside in relatively low numbers within tissues, making isolation difficult [93]. Mast cells leave the bone marrow as progenitors and migrate to peripheral sites where they complete their differentiation under the influence of factors present at each site [94]. Mast cells normally reside in the connective tissue close to epithelia, associated with blood vessels, nerves, and are found in mucosa of the gastrointestinal, respiratory tracts, in the skin, near muscle cells and mucus-producing glands. In some species, including murine rodents, mast cells also occur within mesothelium-lined cavities, such as the peritoneal cavity [95, 96].

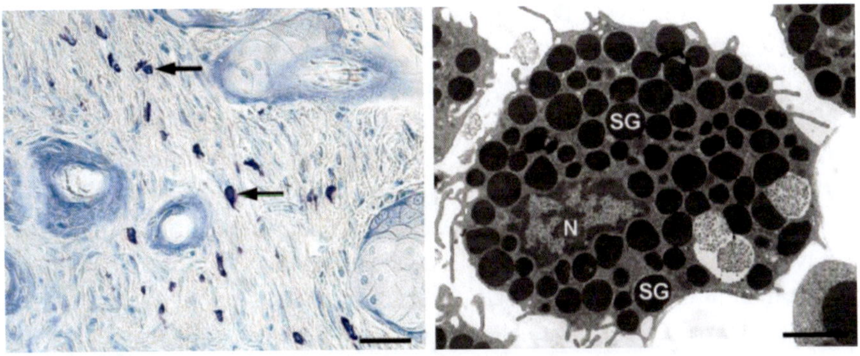

Figure 2. Mast cells in the dermis (arrows) stain metachromatically with toluidine blue. Bar equals 40μm. Transmission electron microscopy. Electron dense secretory granules can be seen in the cytoplasm in this mature peritoneal mast cell. Bar equals 5μm; N, nucleus; SG, secretory granule.

Kitamura and collaborators [97] characterized WBxC57 BL/6-FI-c-kit $^{w/Wv}$ mice (W/W^{v} mice), which are mast cell-deficient. Interestingly, an attractive feature of these mice was that their mast cell deficiency could be reconstituted by adoptive transfer of cultured mast cells. Furthermore, transfer of mast cells with targeted deletions in specific genes allowed the study of mast cell specific expression of particular proteins in a biological response. This model was a useful tool for investigations concerning the physiologic roles and the origin of mast cells [97]. In 2005, another mast cell-deficient mouse, W^{sh}/W^{sh}, was characterized [98]. Similar to W/W^{v} mice, the W^{sh}/W^{sh} mice also have mutations at the *W* locus, however, they are fertile and do not possess some of the c-kit-mediated developmental defects such as anemia, although they do have an age-dependent loss of mast cells that is complete by about 12 weeks of age. Studies using both mouse mast cell deficient models have contributed to a better knowledge of mast cell biology and have revealed that mast cells are multifaceted immune cells with diverse roles in health and disease [93, 99].

Mast cells have long been known to play a critical role in inflammatory and allergic reactions [95, 96, 100-107]. Apart from their roles in pathological conditions, the biological functions of mast cells include roles in both innate and adaptive immunity, involvement in host defense mechanisms against parasites, immunomodulation of the immune system, tissue repair and angiogenesis [100, 101, 108-111]. Recently, mast cells have gained new importance as immunoregulatory cells with the discovery that they are a major source of cytokines and chemokines [93, 96, 112, 113].

Mast cells have a large repertoire of cell-surface receptors, such as Toll-like receptors (TLRs), the high-affinity receptor for immunoglobulin E (FcεRI) and the high-affinity receptor for immunoglobulin G (FcγRI), enabling them to interact both directly and indirectly with pathogens and environmental toxins. Activated mast cells possess the ability to produce and release a wide variety of mediators, which serve to activate the immune response or to amplify an existing response [93].

Although mast cells may be activated by a number of stimuli and pathways [114, 115], the major mechanism of activation and subsequent degranulation is through the FcεRI, present in the plasma membrane of mast cells, epidermal Langerhans cells, eosinophils, and basophils [116]. FcεRI is expressed as a heterotetrameric structure composed of one α subunit with an extracellular domain that binds immunoglobulin E (IgE), a four transmembrane-spanning β subunit, and two identical disulphide linked γ subunits [3, 117, 118]. The carboxyl terminal cytoplasmic domains of both β and γ subunits contain an immunoreceptor tyrosine-based activation motif (ITAM), common to all multi-subunit immune recognition receptors, that is critical for cell activation. Mast cell activation is initiated by the binding of oligomeric antigens to receptor-bound IgE, which crosslinks FcεRI and results in its aggregation.

The first recognized biochemical event of the cytoplasmic signal transduction cascade, after crosslinking of two or more FcεRI, involves phosphorylation, usually by Lyn, a member of the Src family of tyrosine kinases, of two conserved tyrosine residues within the ITAMs of both β and γ subunits of the receptor. The tyrosine-phosphorylated ITAMs create a novel binding surface that is recognized by additional cytoplasmic signaling molecules, such as the protein tyrosine kinase Syk which binds mainly to the γ subunit, via its tandem Src homology 2 (SH2) domains. This interaction results in an increased kinase activity that rapidly shifts the equilibrium of the cell from a resting state to an activated state. This Syk mediated signal amplification results in a direct or indirect activation of several proteins, LAT, Vav, phospholipase C- γ 1 (PLC- γ1) and PLC- γ2. Finally, downstream activation leads to calcium (Ca^{2+}) release from the endoplasmic reticulum (ER), which is followed by Ca^{2+} influx from outside of cell, causing an increase in the cytosolic free Ca^{2+} concentration [119], activation of other enzymes and adaptors, and rearrangement of the cytoskeleton that culminates in the release of three classes of mediators: (1) *preformed mediators* (stored in secretory granules), such as histamine, heparin, β-hexosaminidase, neutral proteases, acid hydrolases, major basic protein, carboxypeptidases, and some cytokines and growth factors; (2) *newly formed lipid mediators*, such as prostaglandins and leukotrienes; and (3) *newly synthesized m... tors* that include growth factors, cytokines and chemokines [104, 117, 120].

Accumulating evidence suggests that lipid rafts or raft components p... pivotal role in signal transduction via FcεRI in mast cells and that t... organization of various molecules in lipid rafts could modulate many biological processes in these cells.

Lipid Rafts, Mast Cells and Signal Transduction

The first signaling complex convincingly shown to involve lipid rafts was IgE during an allergic immune response [88, 121]. IgE signaling was initially thought to be based on protein-protein interactions alone, but several observations indicated that lipid rafts, are involved in this process [20, 88, 122-127]. The first hint of lipid raft involvement came from the finding that FcεRI is soluble in Triton X-100 at steady state, but after crosslinking becomes insoluble in low concentrations of this detergent [88]. Moreover, in unstimulated cells, FcεRI is dispersed throughout the plasma membrane, but upon activation rapidly aggregates [128, 129] and can then be found on the cell surface in association with the ganglioside GM_1 [130-133] and the glycosylphosphatidylinositol-linked protein Thy-1 [134]. Fridriksson and collaborators [135], using Fourier transform mass spectrometry (FTMS) with electrospray ionization mass spectrometry (ESI/MS) compared the phospholipid composition of lipid rafts with plasma membrane vesicles and whole cell lipid extracts from RBL-2H3 cells, a rat mast cell line, and observed that antigen stimulation of FcεRI causes a significant change in the phospholipid composition of lipid rafts, with a substantial increase in the ratio of polyunsaturated to saturated plus monounsaturated phospholipids.

Despite numerous studies on mast cell activation through FcεRI, the detailed mechanism by which cross-linking promotes the initial phosphorylation by Lyn and how and/or where these proteins interact remains unclear [14, 20, 87, 136]. Davey and collaborators [20] suggested that protein-protein interaction (IgE-FcεRI cross-linking) recruits essential signaling proteins and lipid molecules into more ordered domains that serve as a platform for signaling. The structural basis for the association of FcεRI with lipid rafts is partially understood. Both β and γ subunits are palmitoylated which could facilitate their association with lipid rafts [137]. However, Field and collaborators [52] investigated the structural basis for the interaction of FcεRI with isolated lipid rafts and observed that this interaction is not mediated by the β subunit or by the cytoplasmic tail of the γ subunit of the receptor. In this study, the results point to the transmembrane segments of α or γ subunits as being primarily responsible for aggregation-dependent association of FcεRI with lipid rafts. However, more extensive analyses are required to completely define the structural basis for FcεRI and lipid raft association.

An approach extensively used to better understand the role of lipid rafts in FcεRI-mediated signaling has been the study and/or the manipulation of the lipid constituents of rafts, such as cholesterol and gangliosides [3, 138, 139] or electron microscopy studies [66]. Lipid rafts present three to five times more cholesterol than the rest of the plasma membrane, which presumably occurs because the cholesterol has higher affinity for raft sphingolipids than for unsaturated phospholipids. Cholesterol is thought to play role as a spacer between the hydrocarbon chains of the sphingolipids and function as dynamic glue that keeps the raft assembly together [3]. Removal of raft cholesterol leads to dissociation of most proteins from rafts and renders them nonfunctional [140].

Methyl-β-cyclodextrin (MβCD), a carbohydrate molecule with a pocket for binding cholesterol, has been extensively used to deplete the plasma membrane of cholesterol and subsequently disrupt lipid rafts [3]. Also, MβCD has been used to study the role of lipid rafts in FcεRI-mediated signaling, particularly in early events of the signal transduction such as tyrosine phosphorylation of FcεRI by Lyn [138]. Sheets and collaborators [121] demonstrated that phosphorylation of FcεRI occurs in a cholesterol-dependent manner and that cholesterol depletion reduces tyrosine phosphorylation of FcεRI after cross-linking. In parallel to its inhibition of tyrosine phosphorylation, cholesterol depletion disrupts the interactions of aggregated FcεRI and Lyn in intact cells. Cholesterol repletion restores receptor tyrosine phosphorylation together with the structural interactions among the raft constitutes, providing strong evidence that lipid raft structure, maintained by cholesterol, plays a critical role in the initiation of FcεRI signaling. Cholesterol depletion by MβCD in RBL-2H3 cells also reduced the release of β-hexosaminidase activity in cells stimulated via FcεRI [139, 141-143]. These data suggest that cholesterol depletion by MβCD affects IgE signaling due to the disruption of lipid rafts resulting in a failure to form a signaling complex. Additionally, Young and collaborators [87] provided evidence that when Lyn is segregated in lipid rafts it has substantially higher kinase activity than Lyn outside the membrane micro-domains. These data suggest that some undefined components of lipid rafts may influence the kinase activity of Lyn [138] and consequently FcεRI signal transduction.

Another constituent of lipid rafts, f flotillin-1, was initially identified as a caveolae-associated membrane protein and is a marker protein of lipid rafts although its physiolological role is still not clear [144, 145]. Kato and collaborators [138] using flotillin-1 knockdown RBL-2H3 cells showed that flotillin-1 regulates the kinase activity of Lyn in mast cells. In the flotillin-1 knockdown cells there was a significant decrease in Ca^{2+} mobilization, the phosphorylation of ERKs, tyrosine phosphorylation of the γ-subunit of FcεRI, and IgE-mediated degranulation. This study also showed that flotillin-1 is constitutively associated with Lyn in lipid rafts in RBL-2H3 cells, and that antigen stimulation induced an increase in flotillin-1 binding to Lyn, resulting in an enhancement of Lyn kinase activity. These data suggest that flotillin-1 is an important component of FcεRI-mediated mast cell activation, and regulates the kinase activity of Lyn in lipid rafts.

Gangliosides are structurally important for lipid raft assembly and function. The rigid structural nature of the ceramide anchor in gangliosides, coupled with the ability of sphingolipids to associate with cholesterol, is thought to drive the assembly of lipid rafts [3, 146]. Two unique gangliosides (α-galactosyl derivatives of the ganglioside GD_{1b} antigens I and II) are present on the surface of rodent mast cells [147-150] and are specifically recognized by the monoclonal antibody (mAb) AA4. Previous studies using mAb AA4 [148, 149] demonstrated that the α-galactosyl derivatives of the ganglioside GD_{1b} were present only in mast cells and not in any other cell type in all 19 rat [149] and 18 mouse [150] tissues examined, which confirms the specificity of mAb AA4 for rodent mast cells. These α-galactosyl derivatives of the ganglioside GD_{1b}, antigens I and II, contain respectively one and two additional α-galactosyl residues when compared with GD_{1b}. The gangliosides derived from GD_{1b} have been identified as components of lipid rafts in the plasma membrane of RBL-2H3 cells [120, 121, 139]. They also seem to be intimately involved with signaling through FcεRI. Although the functional role of these gangliosides is not clear, previous studies have shown that when the α-galactosyl derivatives of ganglioside GD_{1b} were bound to mAb AA4, histamine release was inhibited in a time and concentration dependent manner [149, 151]. Binding of mAb AA4 to RBL-2H3 cells resulted in an increase in intracellular calcium, phosphatidylinositol hydrolysis, and a redistribution of PKC. However, the magnitude of these changes was less than those after FcεRI cross-linking, and unlike FcεRI activation, these changes were not accompanied by histamine release [149]. These derivatives of the ganglioside GD_{1b} co-precipitated with the Lyn. Although mAb AA4 binds to sites close to FcεRI the association between Lyn and these gangliosides was independent of the association of the gangliosides with FcεRI. Moreover, the association of Lyn with the gangliosides is much stronger than the association of Lyn with FcεRI. These associations suggest that a complex that includes gangliosides, FcεRI and Lyn is essential for modulation of signal transduction in mast cells [149, 151-154].

Furthermore, analysis of the subcellular distribution by sucrose gradients showed that following FcεRI activation there was a shift in the distribution of the gangliosides and FcεRI to the lipid raft fractions [120, 139]. The movement of these gangliosides and of the FcεRI into the lipid rafts may be another mechanism that regulates signal transduction in mast cells by facilitating formation of a signaling complex. In addition, when FcεRI is crosslinked, the raft components patch selectively into restricted cell surface domains [134]. Non-raft proteins are not recruited into these domains [57]. Harder et al. also showed that this co-redistribution of raft components is dependent on cholesterol. These raft domains can selectively incorporate or exclude proteins, and thereby control protein–protein and protein–lipid interactions.

In 2008, Silveira e Souza and collaborators [139], using a cell line deficient in the α-galactosyl derivatives of the ganglioside GD_{1b}, as well as the parent cell line, RBL-2H3, demonstrated the importance of these gangliosides for lipid raft organization and consequently for FcεRI-mediated degranulation in rodent mast cells. In this study the authors observed an increased inhibition in the release of β-hexosaminidase activity in the mutant cell line after FcεRI stimulation, but not after exposure to calcium ionophore. Moreover, treatment of RBL-2H3 cells with compounds that inhibit ganglioside synthesis also reduced release of β-hexosaminidase activity showing that release of β-hexosaminidase activity is dependent on the presence of gangliosides.

In addition to lipid raft assembly, another possible role for the mast cell specific gangliosides in signal transduction may be to facilitate the association between Lyn and FcεRI. Because FcεRI itself has no intrinsic kinase activity, the tyrosine phosphorylation induced by receptor cross-linking is a secondary event that occurs after aggregation of FcεRI and its movement into lipid rafts [154]. Therefore, formation of lipid raft complexes that include gangliosides, and associated proteins, such as Lyn, LAT, flotillin-1 and FcεRI, have an important role in receptor-mediated signal transduction.

In addition to gangliosides, sphingomyelin, a type of sphingolipid also has a role in signaling events in mast cells. The sphingomyelin pathways with its products ceramide, sphingosine (S) and S-1-phosphate (S1P) are known to play roles in signal transduction [155, 156]. Sphingosine kinase has been recognized as an essential signaling molecule that mediates the intracellular conversion of S to S1P. In mast cells, induction of sphingosine kinase and generation of S1P have been linked to the initial rise in Ca^{2+}, released from internal stores, and to degranulation. Urtz and collaborators [157] showed that sphingosine kinase type 1 (SPHK1) interacts directly with Lyn and this interactions leads to the recruitment of this lipid kinase to the FcεRI. The interaction of SPHK1 with Lyn caused enhanced lipid and tyrosine kinase activity. After FcεRI triggering, enhanced sphingosine kinase activity was associated with FcεRI in lipid rafts in mast cells.

Mast cells also produce structures such as cytonemes or tunneling nanotubes used for intercellular communication and this intercellular communication may be important during allergic and inflammatory responses following co-stimulation of FcεRI and CCR1 (chemokine receptor 1) [99]. Albeit the process of cytoneme formation remains poorly understood, the fact that cholesterol depletion reduced the formation of cytonemes, suggests that lipid rafts may participate in cytoneme formation in mast cells, either by promoting membrane integrity or by participating in cell signaling.

Studies have shown that establishing and maintaining lipid rafts are necessary for many biological processes besides cell signaling [158, 159]. In mast cells, the structure-function relationship of lipid rafts or of rafts constitutes, are important in various biological events.

Lipid Rafts and Mast Cells: Morphology, Endocytosis and Cell Development and Recruitment

Morphology

Using the parent cell line, RBL-2H3, [100, 149, 160-162], and a ganglioside deficient cell line, D1, the influence of lipid rafts on cell structure and organization has been examined [163]. The D1 cell line is deficient in GM_1 gangliosides and in the α-galactosyl derivatives of the ganglioside GD_{1b}.

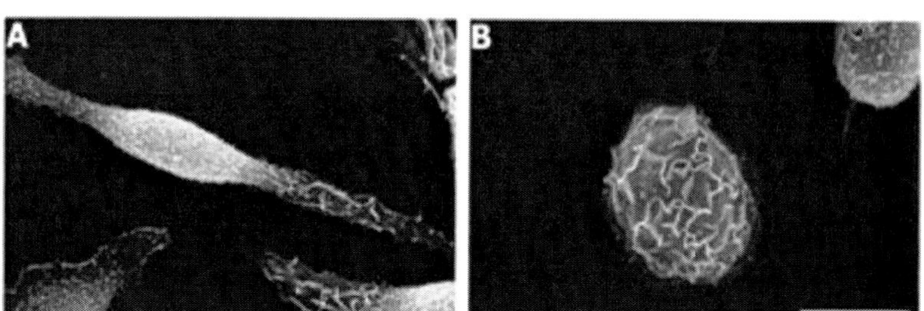

Figure 3. Ganglioside-deficient D1 cells have an altered morphology. By scanning electron microscopy, RBL-2H3 cells (A) are spindle shaped and their surface is covered with short microvilli. In contrast, D1 cells (B) are rounded and their surface is covered with large membrane ruffles. Both cell lines were stimulated for 5 minutes via FcεRI. Bar equals 10μm.

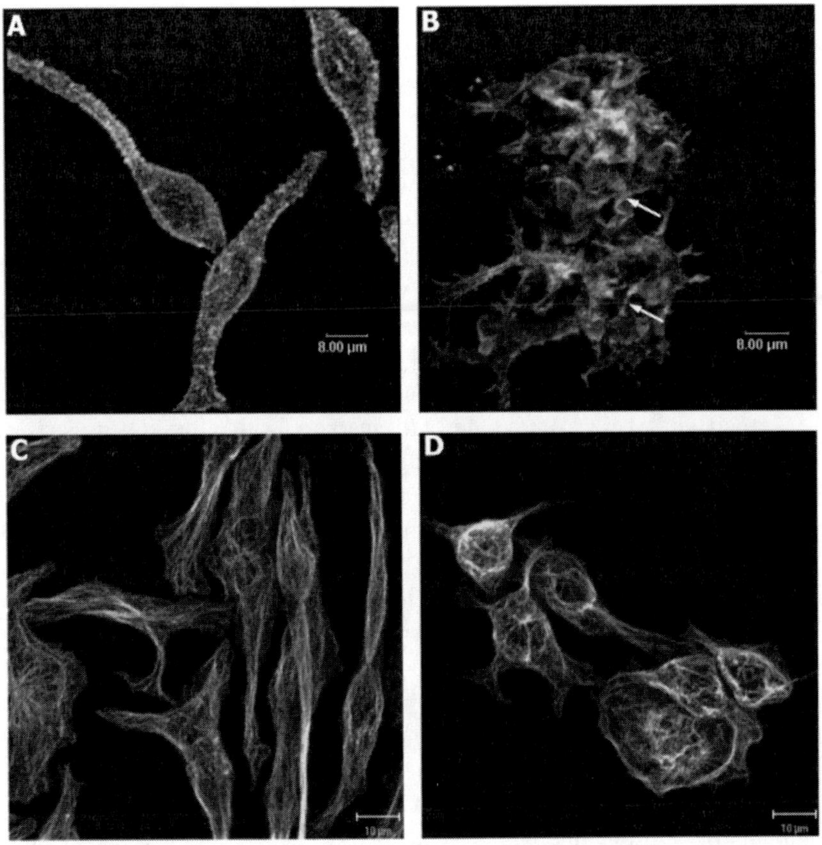

Figure 4. The cytoskeleton is altered in D1 cells. The F-actin distribution in RBL-2H3 and D1 cells reflects their morphology. Actin filaments in RBL-2H3 cells (A) lie under the plasma membrane following the spindle shape of the cells and in association with microvilli. In D1 cells (B), actin filaments are concentrated in large membrane ruffles (arrows) and extend into the cytoplasmic projections. F-actin was labeled with phaloidin-Alexa 488. In RBL-2H3 cells (C), microtubules are organized into bundles that followed the fusiform shape of the cells, whereas in D1 cells (D), microtubules appear as thick bundles that are restricted to the cell body and are not present in the cytoplasmic extensions. To visualize the microtubules, the cells were fixed, permeabilized and labeled with anti-β-tubulin antibody and subsequently with a secondary antibody conjugated to Alexa-488. Samples were examined using a Leica TCS-NT laser scanning confocal microscope.

The mutant cell line D1 has a cellular morphology which is distinct from RBL-2H3 cells. RBL-2H3 are spindle shaped and their surface is covered with short microvilli, while D1 cells are rounded and their surface is covered with large membrane ruffles (Figure 3), suggesting that the gangliosides are important in the maintenance of normal cell morphology.

The morphological changes observed in D1 cells may be related to the lipid composition of these cells. This cell line presents a large decrease in glycosphingolipids, such as GM_1 and the α-galactosyl derivatives of the ganglioside GD_{1b}, which may affect many physicochemical properties of the plasma membrane. According to Kato [142] the lipid composition can influence membrane stability, membrane fluidity, lipid packing, bilayer curvature, and hydration elasticity, as well as anchorage of the cytoskeleton to the plasma membrane. In fact, D1 cells show an abnormal distribution of actin filaments and microtubules (Figure 4) [163]. A growing body of evidence indicates that lipid rafts are essential for membrane-cytoskeleton coupling and the association of Lyn and other raft markers with crosslinked FcεRI is regulated by interactions with F-actin [5, 134, 164]. It is possible that in the D1 mutant cells, the disorganization of both lipid rafts and actin filaments contributes to impaired degranulation after FcεRI stimulation [139, 163]. Furthermore, the actin cytoskeleton is known to participate in regulating and activating raft-associated signaling events [165-167].

The factors that govern the formation of lipid rafts continue to be elucidated, but lipid raft formation requires actin polymerization. The connection of lipid raft proteins with actin filaments directs the lateral distribution and mobility of these membrane proteins [77, 168]. The extent to which the actin cytoskeleton participates in the formation of membrane rafts is not yet established. Han [59] observed, with mass spectrometric analyses, that inhibitors of actin polymerization (cytochalasin D and latrunculin A) cause a similar, but faster, change in phospholipid composition of lipid rafts than activation through FcεRI. Thus, perturbations in the actin filaments affect the organization of lipid rafts in RBL-2H3 cells. The changes in raft phospholipid composition following stimulation depend on actin polymerization and appear largely independent of the tyrosine phosphorylation cascade that is activated in parallel to these cytoskeletal changes.

The actin cytoskeleton is a dynamic structure that changes in response to extracellular signals, and therefore represents one mechanism governing the establishment and distribution of lipid rafts in the plasma membrane [169]. Lipid rafts may be structured by a synergistic interaction between the cortical actin filaments and the lipid rafts themselves, and that many of the structural and functional properties of rafts require an intact actin cytoskeleton [170]. An important regulator of membrane-cytoskeleton interactions is the phosphoinositide PIP2, which is a minor lipid component of the plasma membrane that is known to regulate the organization of the actin cytoskeleton and in particular the formation of actin-membrane linkages [171]. PIP2 also serves as a co-factor for many of the proteins that anchor actin filaments to the plasma membrane [171, 172]. Protein binding to PIP2 often occurs through a PIP2-specific recognition sequence, in many cases represented by a PIP2-specific pleckstrin homology (PH) domain [173-175]. Some actin binding proteins (ABPs) are thought to link actin filaments and PIP2-enriched rafts. Gelsolin is one of the ABPs present in lipid rafts [176].

Microtubules are another of the major determinants of cell shape and polarity [177, 178]. In the ganglioside deficient D1 cells the arrangement of microtubules was completely disorganized demonstrating that the abnormal morphology observed in the mutant cell line could be correlated with the decrease in gangliosides that leads to lipid raft disorganization [163]. Therefore, gangliosides are structurally important for lipid raft assembly and function and are critical for maintenance of mast cell morphology.

Endocytosis

Concomitant with the concept of lipid rafts and the lateral movement of proteins in the plasma membrane, it was observed that plasma membrane proteins could suffer a selective reorganization followed by internalization of these proteins [179-181]. Receptor mediated endocytosis, including endocytosis of FcεRI, is a temporally and spatially organized process [120, 182]. Receptor-induced endocytosis falls into to two main categories, clathrin-dependent and clathrin-independent endocytosis. In general both types of receptor mediated endocytosis are dependent on lipid rafts. Clathrin-dependent endocytosis is the best-studied of the two. This pathway requires the presence of the protein clathrin, which is the major component of the endocytic vesicle coat [183-185]. Cargo associated adapters recruit clathrin to the plasma membrane and promotes its assembly into a polyhedral lattice leading to the formation of a clathrin-coated vesicle. The major adapter for clathrin is the adapter protein complex-2 (AP-2) [200]. Clathrin-independent endocytosis is not as well understood. The major form of clathrin-independent endocytosis is dependent on lipid rafts that contain the protein caveolin. Caveolin induces the formation of caveolae at the cell surface [186, 187]. The formation of a bud followed by pinching-off of the vesicles, to into of cytoplasm, requires the activity of the GTPase dynamin [184, 188]. When endocytosis is lipid raft-dependent, internalization of the receptor is sensitive to cholesterol depletion and the internalized receptor colocalizes with lipid raft components including flotillin and gangliosides such as GM_1 and GD_{1b} [120, 189].

In mast cells, after activation, crosslinked FcεRI is endocytosed through clathrin-coated vesicles and transported by the endosomal system for eventual degradation in lysosomes or dissociation from the ligands [190-192]. Thus, clathrin-dependent endocytosis has long been thought to be a mechanism for the attenuation of receptor-activated responses [119]. In unstimulated mast cells, FcεRIs are dispersed throughout the plasma membrane but upon activation the receptors rapidly aggregate and can be found on the cell surface in lipid rafts in association with GM_1 [129, 133], the α-galactosyl derivatives of the ganglioside GD_{1b}, protein tyrosine kinase Lyn, and LAT [52, 120, 121]. However, only when the mast cells are activated via FcεRI does a significant internalization of the GD_{1b} derivatives occur [120, 193]. Oliver and collaborators [120] showed that upon activation of FcεRI, the gangliosides derived from GD_{1b} are internalized together with the receptor, following the same pathway to lysosomes. In order to verify the importance of the gangliosides derived from GD_{1b}, present in lipid rafts, in endocytosis Mazucato and collaborators [193] used mutant RBL-2H3 cells that were deficient in these gangliosides. The ganglioside-deficient mast cells showed a delay in the internalization of FcεRI into the endocytic compartments, indicating that the gangliosides derived from GD_{1b} are essential for the endocytosis of FcεRI. The endocytic process itself may play an important role in signal transduction [182, 186, 194] and may facilitate the structural preservation of signaling complexes, and the prolongation of the signal since these gangliosides as well as the FcεRI are associated in lipid rafts and remain associated during endocytosis.

In view of the importance of lipid raft integrity for efficient receptor endocytosis, it has been observed that the FcεRI ubiquitination is a key mechanism for the regulation and control of antigen-dependent endocytosis of receptor complexes [189]. Moreover, it has been demonstrated that ubiquitin ligases Cbl and Nedd4 are recruited into lipid rafts upon IgE triggered cell signaling [195]. Nedd4 is known to ubiquitinate membrane receptors [196]. The ubiquitin Cbl is a good candidate to mediate FcεRI ubiquitination since it participates in various functions such as cis-and-trans-ubiquitination [197]. It is phosphorylated upon FcεRI engagement [198] and negatively regulates Syk kinase [199]. Molfetta and collaborators [189, 200] suggested that the recruitment of engaged FcεRI subunits into lipid rafts precedes their ubiquitination, and that integrity of lipid rafts is required for receptor ubiquitination and endocytosis, contributing to the down-regulation of FcεRI-mediated signaling.

Mast Cell Maturation

The expression of lipid raft components, such as the α-galactosyl derivatives of the ganglioside GD_{1b}, on the mast cell surface also appears to be related to mast cell maturation. Previous studies using mAb AA4 showed that the α-galactosyl derivatives of the ganglioside GD_{1b} were present only in mast cells and not in any other cell type in all rat tissues examined [148, 149]. However, in bone marrow, a population of large, poorly differentiated cells, presumably immature mast cells also stained with mAb AA4 [149]. Later these cells were indeed shown to be very immature and immature mast cells [201, 202]. Since the heterogeneity of the maturing mast cells makes them impossible to separate from other cells on the basis of their density and mAb AA4 binds only to cells which can be identified as mast cells [201, 203], the gangliosides recognized by mAb AA4 may be considered a powerful marker for rodent mast cells. On the other hand, undifferentiated mast cell precursors in the bone marrow do not express, the α-galactosyl derivatives of the ganglioside GD_{1b} recognized by mAb AA4. Jamur and collaborators [150] isolated and characterized a committed mast cell precursor in the bone marrow of adult Balb/c mice, by immunoaffinity selection using 2 monoclonal antibodies, mAb AA4 and mAb BGD6. While mAb AA4 recognizes 2 derivatives of the ganglioside GD_{1b} that are unique to the surface of rodent mast cells [147-149], mAb BGD6 binds to the surface of RBL-2H3 mast cells at sites unrelated to FcεRI. Both mAb AA4 and mAb BGD6 bind to granulated mast cells in all stages of maturation in the bone marrow as well as binding to a very immature cell that is not recognized by mAb AA4. This mast cell precursor was characterized as BGD6+, $CD34^+$, $CD13^+$, c-kit^+, $FcεRI^-$ and $AA4^-$ (negative for the expression of derivatives of the ganglioside GD_{1b}). These gangliosides begin to be expressed on the cell surface jointly with FcεRI and at the same time as the initiation of the formation of cytoplasmic granules in very immature mast cells and continue to be expressed by mast cells in all stages of maturation [150]. These data suggest that mast cell lipid rafts or raft constitutes are related to mast cell maturation and function.

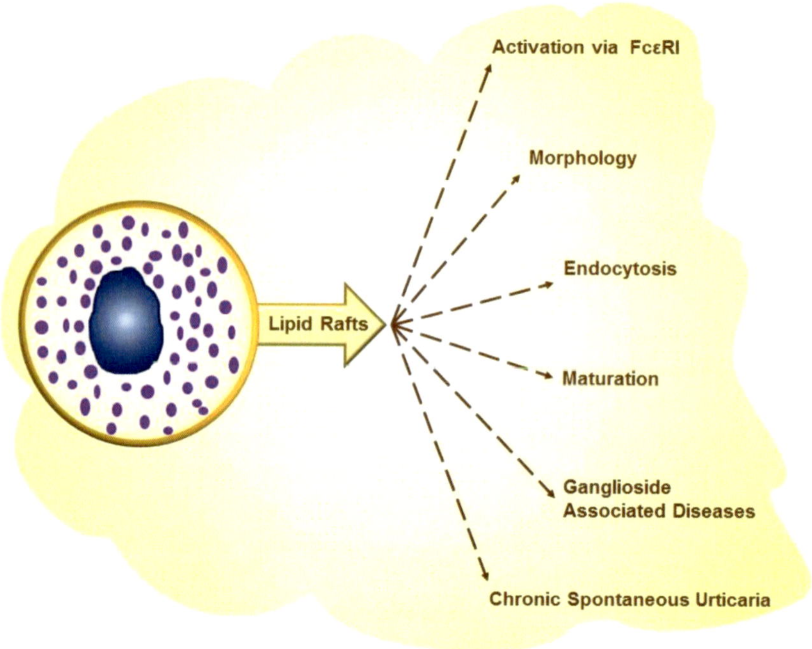

Figure 5. The diagram represents the involvement of lipid rafts in various aspects of mast cell biology. Lipid rafts are important in the modulation of many cellular processes such as activation through FcεRI, maintenance of cell structure, endocytosis, maturation, and are relevant in diseases related to gangliosides.

Conclusion

Although several aspects of lipid raft structure and function in mast cells still remain to be elucidated, this chapter presents evidence that, undoubtedly, lipid rafts and their constitutes modulate many of biological processes in mast cells (Figure 5). Further research to better define the role of lipid rafts in mast cells could offer novel targets for immunotherapies and treatment of diseases in which mast cells and/or their mediators are involved.

Acknowledgments

We would like to thank CAPES, CNPq, FAEPA, and FAPESP for financial support.

References

[1] Singer SJ, Nicolson GL. The fluid mosaic model of the structure of cell membranes. *Science*. 1972 Feb 18;175(23):720-31.

[2] Brown DA, London E. Functions of lipid rafts in biological membranes. *Annu. Rev. Cell Dev. Biol.* 1998;14:111-36.

[3] Simons K, Toomre D. Lipid rafts and signal transduction. *Nat. Rev. Mol. Cell Biol.* 2000 Oct;1(1):31-9.

[4] Harder T. Raft membrane domains and immunoreceptor functions. *Adv. Immunol.* 2001;77:45-92.

[5] Pike LJ. The challenge of lipid rafts. *J. Lipid Res.* 2009 Apr;50 Suppl:S323-8.

[6] Lee AG, Birdsall NJ, Metcalfe JC, Toon PA, Warren GB. Clusters in lipid bilayers and the interpretation of thermal effects in biological membranes. *Biochemistry*. 1974 Aug 27;13(18):3699-705.

[7] Wunderlich F, Giese G, Bucherer C. Expansion and apparent fluidity decrease of nuclear membranes induced by low Ca/Mg. Modulation of nuclear membrane lipid fluidity by the membrane-associated nuclear matrix proteins? *J. Cell Biol.* 1978 Nov;79(2 Pt 1):479-90.

[8] Karnovsky MJ, Kleinfeld AM, Hoover RL, Klausner RD. The concept of lipid domains in membranes. *J. Cell Biol.* 1982 Jul;94(1):1-6.

[9] Dolo V, D'Ascenzo S, Sorice M, Pavan A, Sciannamblo M, Prinetti A, et al. New approaches to the study of sphingolipid enriched membrane domains: the use of electron microscopic autoradiography to reveal metabolically tritium labeled sphingolipids in cell cultures. *Glycoconjugate journal*. 2000 Mar-Apr;17(3 -4):261-8.

[10] Thompson TE, Tillack TW. Organization of glycosphingolipids in bilayers and plasma membranes of mammalian cells. *Annu. Rev. Biophys. Biophys. Chem.* 1985;14:361-86.

[11] Smart EJ, Ying YS, Mineo C, Anderson RG. A detergent-free method for purifying caveolae membrane from tissue culture cells. *Proc. Natl. Acad. Sci. U. S. A.* 1995 Oct 24;92(22):10104-8.

[12] Rodgers W, Rose JK. Exclusion of CD45 inhibits activity of p56lck associated with glycolipid-enriched membrane domains. *The Journal of cell biology*. 1996 Dec;135(6 Pt 1):1515-23.

[13] Iwabuchi K, Yamamura S, Prinetti A, Handa K, Hakomori S. GM3-enriched microdomain involved in cell adhesion and signal transduction

through carbohydrate-carbohydrate interaction in mouse melanoma B16 cells. *The Journal of biological chemistry*. 1998 Apr 10;273(15):9130-8.
[14] Draber P, Draberova L. Lipid rafts in mast cell signaling. *Mol. Immunol.* 2002 Sep;38(16-18):1247-52.
[15] Pike LJ, Han X, Chung KN, Gross RW. Lipid rafts are enriched in arachidonic acid and plasmenylethanolamine and their composition is independent of caveolin-1 expression: a quantitative electrospray ionization/mass spectrometric analysis. *Biochemistry*. 2002 Feb 12;41(6): 2075-88.
[16] Suzuki KG. Lipid rafts generate digital-like signal transduction in cell plasma membranes. *Biotechnology journal*. Jun;7(6):753-61.
[17] George KS, Wu S. Lipid raft: A floating island of death or survival. *Toxicology and applied pharmacology*. Mar 15;259(3):311-9.
[18] Lemaire-Ewing S, Lagrost L, Neel D. Lipid rafts: a signalling platform linking lipoprotein metabolism to atherogenesis. *Atherosclerosis*. Apr;221(2):303-10.
[19] Szoor A, Szollosi J, Vereb G. Rafts and the battleships of defense: the multifaceted microdomains for positive and negative signals in immune cells. *Immunology letters*. 2010 May 4;130(1-2):2-12.
[20] Davey AM, Walvick RP, Liu Y, Heikal AA, Sheets ED. Membrane order and molecular dynamics associated with IgE receptor cross-linking in mast cells. *Biophys. J.* 2007 Jan 1;92(1):343-55.
[21] Brugger B, Graham C, Leibrecht I, Mombelli E, Jen A, Wieland F, et al. The membrane domains occupied by glycosylphosphatidylinositol-anchored prion protein and Thy-1 differ in lipid composition. *The Journal of biological chemistry*. 2004 Feb 27;279(9):7530-6.
[22] Hancock JF. Lipid rafts: contentious only from simplistic standpoints. *Nat. Rev. Mol. Cell Biol.* 2006 Jun;7(6):456-62.
[23] Pike LJ. Rafts defined: a report on the Keystone Symposium on Lipid Rafts and Cell Function. *Journal of lipid research*. 2006 Jul;47(7):1597-8.
[24] Lingwood D, Simons K. Lipid rafts as a membrane-organizing principle. Science 2010 Jan 1 [cited 327 5961]; 46-50]. Available from: http://www.ncbi.nlm.nih.gov/entrez/query.fcgi?cmd=Retrieve&db=PubMed&dopt=Citation&list_uids=20044567
[25] Suzuki KG. Lipid rafts generate digital-like signal transduction in cell plasma membranes. *Biotechnol. J.* 2012 Jun;7(6):753-61.

[26] Simons K, Ikonen E. Functional rafts in cell membranes. *Nature*. 1997 Jun 5;387(6633):569-72.

[27] Melkonian KA, Ostermeyer AG, Chen JZ, Roth MG, Brown DA. Role of lipid modifications in targeting proteins to detergent-resistant membrane rafts. Many raft proteins are acylated, while few are prenylated. *The Journal of biological chemistry*. 1999 Feb 5;274(6): 3910-7.

[28] Pike LJ. Lipid rafts: bringing order to chaos. *Journal of lipid research*. 2003 Apr;44(4):655-67.

[29] Rajendran L, Simons K. Lipid rafts and membrane dynamics. *J. Cell Sci.* 2005 Mar 15;118(Pt 6):1099-102.

[30] Szoor A, Szollosi J, Vereb G. Rafts and the battleships of defense: the multifaceted microdomains for positive and negative signals in immune cells. *Immunology letters*. 2010 May 4;130(1-2):2-12.

[31] Yamabhai M, Anderson RG. Second cysteine-rich region of epidermal growth factor receptor contains targeting information for caveolae/rafts. *The Journal of biological chemistry*. 2002 Jul 12;277(28):24843-6.

[32] Scheiffele P, Roth MG, Simons K. Interaction of influenza virus haemagglutinin with sphingolipid-cholesterol membrane domains via its transmembrane domain. *The EMBO journal*. 1997 Sep 15;16(18):5501-8.

[33] Crossthwaite AJ, Seebacher T, Masada N, Ciruela A, Dufraux K, Schultz JE, et al. The cytosolic domains of Ca2+-sensitive adenylyl cyclases dictate their targeting to plasma membrane lipid rafts. *The Journal of biological chemistry*. 2005 Feb 25;280(8):6380-91.

[34] Niemela PS, Hyvonen MT, Vattulainen I. Influence of chain length and unsaturation on sphingomyelin bilayers. *Biophysical journal*. 2006 Feb 1;90(3):851-63.

[35] Gidwani A, Brown HA, Holowka D, Baird B. Disruption of lipid order by short-chain ceramides correlates with inhibition of phospholipase D and downstream signaling by FcepsilonRI. *Journal of cell science*. 2003 Aug 1;116(Pt 15):3177-87.

[36] Simons K, Vaz WL. Model systems, lipid rafts, and cell membranes. *Annual review of biophysics and biomolecular structure*. 2004;33:269-95.

[37] Li XM, Momsen MM, Smaby JM, Brockman HL, Brown RE. Cholesterol decreases the interfacial elasticity and detergent solubility of sphingomyelins. *Biochemistry*. 2001 May 22;40(20):5954-63.
[38] Brown DA, Rose JK. Sorting of GPI-anchored proteins to glycolipid-enriched membrane subdomains during transport to the apical cell surface. *Cell*. 1992 Feb 7;68(3):533-44.
[39] Simons K, Gerl MJ. Revitalizing membrane rafts: new tools and insights. *Nat. Rev. Mol. Cell Biol*. 2010 Oct;11(10):688-99.
[40] Lemaire-Ewing S, Lagrost L, Neel D. Lipid rafts: a signalling platform linking lipoprotein metabolism to atherogenesis. *Atherosclerosis*. 2012 Apr;221(2):303-10.
[41] Ikonen E. Roles of lipid rafts in membrane transport. *Current opinion in cell biology*. 2001 Aug;13(4):470-7.
[42] Edidin M. The state of lipid rafts: from model membranes to cells. *Annual review of biophysics and biomolecular structure*. 2003;32:257-83.
[43] Munro S. Lipid rafts: elusive or illusive? *Cell*. 2003 Nov 14;115(4):377-88.
[44] Edidin M. Lipids on the frontier: a century of cell-membrane bilayers. *Nature reviews*. 2003 May;4(5):414-8.
[45] Silveira ESAM, Mazucato VM, Jamur MC, Oliver C. Lipid rafts in mast cell biology. *J. Lipids*. 2011;2011:752906.
[46] Calder PC, Yaqoob P. Lipid rafts--composition, characterization, and controversies. *The Journal of nutrition*. 2007 Mar;137(3):545-7.
[47] Vereb G, Szollosi J, Matko J, Nagy P, Farkas T, Vigh L, et al. Dynamic, yet structured: The cell membrane three decades after the Singer-Nicolson model. *Proc. Natl. Acad. Sci. U. S. A*. 2003 Jul 8;100(14): 8053-8.
[48] Kaiser HJ, Lingwood D, Levental I, Sampaio JL, Kalvodova L, Rajendran L, et al. Order of lipid phases in model and plasma membranes. *Proceedings of the National Academy of Sciences of the United States of America*. 2009 Sep 29;106(39):16645-50.
[49] Rivera J, Arudchandran R, Gonzalez-Espinosa C, Manetz TS, Xirasagar S. A perspective: regulation of IgE receptor-mediated mast cell responses by a LAT-organized plasma membrane-localized signaling complex. *International archives of allergy and immunology*. 2001 Jan-Mar;124(1-3):137-41.

[50] Kimura T, Hisano M, Inoue Y, Adachi M. Tyrosine phosphorylation of the linker for activator of T cells in mast cells by stimulation with the high affinity IgE receptor. *Immunology letters*. 2001 Jan 1;75(2):123-9.

[51] Mayor S, Maxfield FR. Insolubility and redistribution of GPI-anchored proteins at the cell surface after detergent treatment. *Molecular biology of the cell*. 1995 Jul;6(7):929-44.

[52] Field KA, Holowka D, Baird B. Structural aspects of the association of FcepsilonRI with detergent-resistant membranes. *The Journal of biological chemistry*. 1999 Jan 15;274(3):1753-8.

[53] Parolini I, Topa S, Sorice M, Pace A, Ceddia P, Montesoro E, et al. Phorbol ester-induced disruption of the CD4-Lck complex occurs within a detergent-resistant microdomain of the plasma membrane. Involvement of the translocation of activated protein kinase C isoforms. *The Journal of biological chemistry*. 1999 May 14;274(20):14176-87.

[54] Montixi C, Langlet C, Bernard AM, Thimonier J, Dubois C, Wurbel MA, et al. Engagement of T cell receptor triggers its recruitment to low-density detergent-insoluble membrane domains. *The EMBO journal*. 1998 Sep 15;17(18):5334-48.

[55] Surviladze Z, Draberova L, Kubinova L, Draber P. Functional heterogeneity of Thy-1 membrane microdomains in rat basophilic leukemia cells. *European journal of immunology*. 1998 Jun;28(6):1847-58.

[56] Ilangumaran S, Arni S, van Echten-Deckert G, Borisch B, Hoessli DC. Microdomain-dependent regulation of Lck and Fyn protein-tyrosine kinases in T lymphocyte plasma membranes. *Molecular biology of the cell*. 1999 Apr;10(4):891-905.

[57] Harder T, Kuhn M. Selective accumulation of raft-associated membrane protein LAT in T cell receptor signaling assemblies. *J. Cell Biol*. 2000 Oct 16;151(2):199-208.

[58] Surviladzc Z, Draberova L, Kovarova M, Boubelik M, Draber P. Differential sensitivity to acute cholesterol lowering of activation mediated via the high-affinity IgE receptor and Thy-1 glycoprotein. *European journal of immunology*. 2001 Jan;31(1):1-10.

[59] Han X, Smith NL, Sil D, Holowka DA, McLafferty FW, Baird BA. IgE receptor-mediated alteration of membrane-cytoskeleton interactions revealed by mass spectrometric analysis of detergent-resistant membranes. *Biochemistry*. 2009 Jul 14;48(27):6540-50.

[60] Baumgart T, Hammond AT, Sengupta P, Hess ST, Holowka DA, Baird BA, et al. Large-scale fluid/fluid phase separation of proteins and lipids in giant plasma membrane vesicles. *Proceedings of the National Academy of Sciences of the United States of America*. 2007 Feb 27;104(9):3165-70.

[61] Sengupta P, Hammond A, Holowka D, Baird B. Structural determinants for partitioning of lipids and proteins between coexisting fluid phases in giant plasma membrane vesicles. *Biochimica et biophysica acta*. 2008 Jan;1778(1):20-32.

[62] Lingwood BE, Healy GN, Sullivan SM, Pow DV, Colditz PB. MAP2 provides reliable early assessment of neural injury in the newborn piglet model of birth asphyxia. *Journal of neuroscience methods*. 2008 Jun 15;171(1):140-6.

[63] Johnson SA, Stinson BM, Go MS, Carmona LM, Reminick JI, Fang X, et al. Temperature-dependent phase behavior and protein partitioning in giant plasma membrane vesicles. *Biochimica et biophysica acta*. 2010 Jul;1798(7):1427-35.

[64] Varma R, Mayor S. GPI-anchored proteins are organized in submicron domains at the cell surface. *Nature*. 1998 Aug 20;394(6695):798-801.

[65] Pralle A, Keller P, Florin EL, Simons K, Horber JK. Sphingolipid-cholesterol rafts diffuse as small entities in the plasma membrane of mammalian cells. *The Journal of cell biology*. 2000 Mar 6;148(5):997-1008.

[66] Wilson BS, Pfeiffer JR, Oliver JM. Observing FcepsilonRI signaling from the inside of the mast cell membrane. *The Journal of cell biology*. 2000 May 29;149(5):1131-42.

[67] Parton RG, Richards AA. Lipid rafts and caveolae as portals for endocytosis: new insights and common mechanisms. *Traffic* (Copenhagen, Denmark). 2003 Nov;4(11):724-38.

[68] Biro A, Cervenak L, Balogh A, Lorincz A, Uray K, Horvath A, et al. Novel anti-cholesterol monoclonal immunoglobulin G antibodies as probes and potential modulators of membrane raft-dependent immune functions. *Journal of lipid research*. 2007 Jan;48(1):19-29.

[69] Gombos I, Steinbach G, Pomozi I, Balogh A, Vamosi G, Gansen A, et al. Some new faces of membrane microdomains: a complex confocal fluorescence, differential polarization, and FCS imaging study on live immune cells. *Cytometry* A. 2008 Mar;73(3):220-9.

[70] Coskun U, Simons K. Membrane rafting: from apical sorting to phase segregation. *FEBS letters*. 2010 May 3;584(9):1685-93.

[71] Pike LJ. Lipid rafts: heterogeneity on the high seas. *Biochem. J.* 2004 Mar 1;378(Pt 2):281-92.

[72] Parton RG, Simons K. The multiple faces of caveolae. *Nature reviews*. 2007 Mar;8(3):185-94.

[73] Mukherjee S, Maxfield FR. Membrane domains. *Annual review of cell and developmental biology*. 2004;20:839-66.

[74] Smith AE, Helenius A. How viruses enter animal cells. *Science* (New York, NY. 2004 Apr 9;304(5668):237-42.

[75] Munoz S, Rivas-Santiago B, Enciso JA. Mycobacterium tuberculosis entry into mast cells through cholesterol-rich membrane microdomains. *Scandinavian journal of immunology*. 2009 Sep;70(3):256-63.

[76] George KS, Wu S. Lipid raft: A floating island of death or survival. *Toxicol. Appl. Pharmacol.* 2012 Mar 15;259(3):311-9.

[77] Rodgers W, Glaser M. Distributions of proteins and lipids in the erythrocyte membrane. *Biochemistry*. 1993 Nov 30;32(47):12591-8.

[78] Kabouridis PS, Magee AI, Ley SC. S-acylation of LCK protein tyrosine kinase is essential for its signalling function in T lymphocytes. *The EMBO journal*. 1997 Aug 15;16(16):4983-98.

[79] Zhang W, Trible RP, Samelson LE. LAT palmitoylation: its essential role in membrane microdomain targeting and tyrosine phosphorylation during T cell activation. *Immunity*. 1998 Aug;9(2):239-46.

[80] Xavier R, Brennan T, Li Q, McCormack C, Seed B. Membrane compartmentation is required for efficient T cell activation. *Immunity*. 1998 Jun;8(6):723-32.

[81] Roy S, Luetterforst R, Harding A, Apolloni A, Etheridge M, Stang E, et al. Dominant-negative caveolin inhibits H-Ras function by disrupting cholesterol-rich plasma membrane domains. *Nature cell biology*. 1999 Jun;1(2):98-105.

[82] Viola A, Schroeder S, Sakakibara Y, Lanzavecchia A. T lymphocyte costimulation mediated by reorganization of membrane microdomains. *Science* (New York, NY. 1999 Jan 29;283(5402):680-2.

[83] Brdicka T, Pavlistova D, Leo A, Bruyns E, Korinek V, Angelisova P, et al. Phosphoprotein associated with glycosphingolipid-enriched microdomains (PAG), a novel ubiquitously expressed transmembrane adaptor protein, binds the protein tyrosine kinase csk and is involved in

regulation of T cell activation. *The Journal of experimental medicine*. 2000 May 1;191(9):1591-604.

[84] Kusumi A, Ike H, Nakada C, Murase K, Fujiwara T. Single-molecule tracking of membrane molecules: plasma membrane compartmentalization and dynamic assembly of raft-philic signaling molecules. *Seminars in immunology*. 2005 Feb;17(1):3-21.

[85] Gilfillan AM, Tkaczyk C. Integrated signalling pathways for mast-cell activation. *Nature reviews*. 2006 Mar;6(3):218-30.

[86] Hunter AJ, Ottoson N, Boerth N, Koretzky GA, Shimizu Y. Cutting edge: a novel function for the SLAP-130/FYB adapter protein in beta 1 integrin signaling and T lymphocyte migration. *J. Immunol*. 2000 Feb 1;164(3):1143-7.

[87] Young RM, Holowka D, Baird B. A lipid raft environment enhances Lyn kinase activity by protecting the active site tyrosine from dephosphorylation. *The Journal of biological chemistry*. 2003 Jun 6;278(23):20746-52.

[88] Field KA, Holowka D, Baird B. Fc epsilon RI-mediated recruitment of p53/56lyn to detergent-resistant membrane domains accompanies cellular signaling. *Proc. Natl. Acad. Sci. U. S. A*. 1995 Sep 26;92(20): 9201-5.

[89] Miceli MC, Moran M, Chung CD, Patel VP, Low T, Zinnanti W. Co-stimulation and counter-stimulation: lipid raft clustering controls TCR signaling and functional outcomes. *Seminars in immunology*. 2001 Apr;13(2):115-28.

[90] Cheng PC, Cherukuri A, Dykstra M, Malapati S, Sproul T, Chen MR, et al. Floating the raft hypothesis: the roles of lipid rafts in B cell antigen receptor function. *Seminars in immunology*. 2001 Apr;13(2):107-14.

[91] Holowka D, Baird B. Fc(epsilon)RI as a paradigm for a lipid raft-dependent receptor in hematopoietic cells. *Seminars in immunology*. 2001 Apr;13(2):99-105.

[92] Ehrlich P. Beitra¨ge zur Theorie und Praxis der histologischen Fa¨rbung.; 1878.

[93] Rao KN, Brown MA. Mast cells: multifaceted immune cells with diverse roles in health and disease. *Ann. N. Y. Acad. Sci*. 2008 Nov;1143:83-104.

[94] Jamur MC, Oliver C. Origin, maturation and recruitment of mast cell precursors. *Frontiers in bioscience* (Scholar edition). 2011;3:1390-406.

[95] Robbie-Ryan M, Brown M. The role of mast cells in allergy and autoimmunity. *Curr. Opin. Immunol*. 2002 Dec;14(6):728-33.

[96] Galli SJ, Kalesnikoff J, Grimbaldeston MA, Piliponsky AM, Williams CM, Tsai M. Mast cells as "tunable" effector and immunoregulatory cells: recent advances. *Annual review of immunology*. 2005;23:749-86.

[97] Kitamura Y, Go S, Hatanaka K. Decrease of mast cells in W/Wv mice and their increase by bone marrow transplantation. *Blood*. 1978 Aug;52(2):447-52.

[98] Grimbaldeston MA, Chen CC, Piliponsky AM, Tsai M, Tam SY, Galli SJ. Mast cell-deficient W-sash c-kit mutant Kit W-sh/W-sh mice as a model for investigating mast cell biology in vivo. *The American journal of pathology*. 2005 Sep;167(3):835-48.

[99] Fifadara NH, Beer F, Ono S, Ono SJ. Interaction between activated chemokine receptor 1 and FcepsilonRI at membrane rafts promotes communication and F-actin-rich cytoneme extensions between mast cells. *International immunology*. 2010 Feb;22(2):113-28.

[100] Metcalfe DD, Baram D, Mekori YA. Mast cells. *Physiol. Rev*. 1997 Oct;77(4):1033-79.

[101] Abraham SN, Malaviya R. Mast cells in infection and immunity. *Infect. Immun*. 1997 Sep;65(9):3501-8.

[102] Ribatti D, Vacca A, Nico B, Crivellato E, Roncali L, Dammacco F. The role of mast cells in tumour angiogenesis. *British journal of haematology*. 2001 Dec;115(3):514-21.

[103] Galli SJ. Mast cells and basophils. *Curr. Opin. Hematol*. 2000 Jan; 7(1):32-9.

[104] Siraganian RP. Mast cell signal transduction from the high-affinity IgE receptor. *Curr. Opin. Immunol*. 2003 Dec;15(6):639-46.

[105] Soucek L, Lawlor ER, Soto D, Shchors K, Swigart LB, Evan GI. Mast cells are required for angiogenesis and macroscopic expansion of Myc-induced pancreatic islet tumors. *Nature medicine*. 2007 Oct;13(10): 1211-8.

[106] Huang B, Lei Z, Zhang GM, Li D, Song C, Li B, et al. SCF-mediated mast cell infiltration and activation exacerbate the inflammation and immunosuppression in tumor microenvironment. *Blood*. 2008 Aug 15;112(4):1269-79.

[107] Wierzbicki M, Brzezinska-Blaszczyk E. [The role of mast cells in the development of inflammatory bowel diseases]. *Postepy higieny i medycyny doswiadczalnej* (Online). 2008;62:642-50.

[108] Garbuzenko E, Nagler A, Pickholtz D, Gillery P, Reich R, Maquart FX, et al. Human mast cells stimulate fibroblast proliferation, collagen synthesis and lattice contraction: a direct role for mast cells in skin fibrosis. *Clin. Exp. Allergy*. 2002 Feb;32(2):237-46.

[109] Maurer M, Theoharides T, Granstein RD, Bischoff SC, Bienenstock J, Henz B, et al. What is the physiological function of mast cells? *Experimental dermatology*. 2003 Dec;12(6):886-910.

[110] Weller K, Foitzik K, Paus R, Syska W, Maurer M. Mast cells are required for normal healing of skin wounds in mice. *Faseb J*. 2006 Nov;20(13):2366-8.

[111] de Souza DA, Jr., Toso VD, Campos MR, Lara VS, Oliver C, Jamur MC. Expression of Mast Cell Proteases Correlates with Mast Cell Maturation and Angiogenesis during Tumor Progression. *PloS one*. 2012;7(7):e40790.

[112] Kalesnikoff J, Galli SJ. New developments in mast cell biology. *Nature immunology*. 2008 Nov;9(11):1215-23.

[113] Beaven MA. Our perception of the mast cell from Paul Ehrlich to now. *European journal of immunology*. 2009 Jan;39(1):11-25.

[114] Crivellato E, Nico B, Mallardi F, Beltrami CA, Ribatti D. Piecemeal degranulation as a general secretory mechanism? *The anatomical record*. 2003 Sep;274(1):778-84.

[115] de Almeida Buranello PA, Moulin MR, Souza DA, Jamur MC, Roque-Barreira MC, Oliver C. The lectin ArtinM induces recruitment of rat mast cells from the bone marrow to the peritoneal cavity. *PloS one*. 2010;5(3):e9776.

[116] von Bubnoff D, Novak N, Kraft S, Bieber T. The central role of FcepsilonRI in allergy. *Clinical and experimental dermatology*. 2003 Mar;28(2):184-7.

[117] Metzger H. The receptor with high affinity for IgE. *Immunol. Rev*. 1992 Feb;125:37-48.

[118] Kinet JP. The high-affinity IgE receptor (Fc epsilon RI): from physiology to pathology. *Annual review of immunology*. 1999;17:931-72.

[119] Ryu SD, Lee HS, Suk HY, Park CS, Choi OH. Cross-linking of FcepsilonRI causes Ca2+ mobilization via a sphingosine kinase pathway in a clathrin-dependent manner. *Cell Calcium*. 2009 Feb;45(2):99-108.

[120] Oliver C, Fujimura A, Silveira ESAM, Orlandini de Castro R, Siraganian RP, Jamur MC. Mast cell-specific gangliosides and FcepsilonRI follow the same endocytic pathway from lipid rafts in RBL-2H3 cells. *J. Histochem. Cytochem*. 2007 Apr;55(4):315-25.

[121] Sheets ED, Holowka D, Baird B. Membrane organization in immunoglobulin E receptor signaling. *Curr. Opin. Chem. Biol.* 1999 Feb;3(1):95-9.

[122] Hanai K, Ishii K, Nojiri H. Sway of the center of gravity in patients with spinal canal stenosis. *Spine* (Phila Pa 1976). 1988 Nov;13(11):1303-7.

[123] Weis FM, Davis RJ. Regulation of epidermal growth factor receptor signal transduction. Role of gangliosides. *The Journal of biological chemistry*. 1990 Jul 15;265(20):12059-66.

[124] Baird B, Sheets ED, Holowka D. How does the plasma membrane participate in cellular signaling by receptors for immunoglobulin E? *Biophysical chemistry*. 1999 Dec 13;82(2-3):109-19.

[125] Metzger H. It's spring, and thoughts turn to...allergies. *Cell*. 1999 Apr 30;97(3):287-90.

[126] Field KA, Apgar JR, Hong-Geller E, Siraganian RP, Baird B, Holowka D. Mutant RBL mast cells defective in Fc epsilon RI signaling and lipid raft biosynthesis are reconstituted by activated Rho-family GTPases. *Molecular biology of the cell*. 2000 Oct;11(10):3661-73.

[127] Wang XQ, Sun P, Paller AS. Ganglioside modulation regulates epithelial cell adhesion and spreading via ganglioside-specific effects on signaling. *The Journal of biological chemistry*. 2002 Oct 25;277(43):40410-9.

[128] Stump RF, Pfeiffer JR, Schneebeck MC, Seagrave JC, Oliver JM. Mapping gold-labeled receptors on cell surfaces by backscattered electron imaging and digital image analysis: studies of the IgE receptor on mast cells. *The American journal of anatomy*. 1989 Jun-Jul;185(2-3):128-41.

[129] Wilson BS, Pfeiffer JR, Surviladze Z, Gaudet EA, Oliver JM. High resolution mapping of mast cell membranes reveals primary and secondary domains of Fc(epsilon)RI and LAT. *The Journal of cell biology*. 2001 Aug 6;154(3):645-58.

[130] Stauffer TP, Meyer T. Compartmentalized IgE receptor-mediated signal transduction in living cells. *The Journal of cell biology*. 1997 Dec 15;139(6):1447-54.

[131] Harder T, Scheiffele P, Verkade P, Simons K. Lipid domain structure of the plasma membrane revealed by patching of membrane components. *The Journal of cell biology*. 1998 May 18;141(4):929-42.

[132] Brown DA, London E. Structure and function of sphingolipid- and cholesterol-rich membrane rafts. *The Journal of biological chemistry*. 2000 Jun 9;275(23):17221-4.

[133] Wilson BS, Steinberg SL, Liederman K, Pfeiffer JR, Surviladze Z, Zhang J, et al. Markers for detergent-resistant lipid rafts occupy distinct and dynamic domains in native membranes. *Molecular biology of the cell*. 2004 Jun;15(6):2580-92.

[134] Holowka D, Sheets ED, Baird B. Interactions between Fc(epsilon)RI and lipid raft components are regulated by the actin cytoskeleton. *Journal of cell science*. 2000 Mar;113 (Pt 6):1009-19.

[135] Fridriksson EK, Baird B, McLafferty FW. Electrospray mass spectra from protein electroeluted from sodium dodecylsulfate polyacrylamide gel electrophoresis gels. *Journal of the American Society for Mass Spectrometry*. 1999 May;10(5):453-5.

[136] Pyenta PS, Holowka D, Baird B. Cross-correlation analysis of inner-leaflet-anchored green fluorescent protein co-redistributed with IgE receptors and outer leaflet lipid raft components. *Biophysical journal*. 2001 May;80(5):2120-32.

[137] Kinet JP, Alcaraz G, Leonard A, Wank S, Metzger H. Dissociation of the receptor for immunoglobulin E in mild detergents. *Biochemistry*. 1985 Jul 16;24(15):4117-24.

[138] Kato N, Nakanishi M, Hirashima N. Flotillin-1 regulates IgE receptor-mediated signaling in rat basophilic leukemia (RBL-2H3) cells. *J. Immunol.* 2006 Jul 1;177(1):147-54.

[139] Silveira e Souza AM, Mazucato VM, de Castro RO, Matioli F, Ciancaglini P, de Paiva Paulino T, et al. The alpha-galactosyl derivatives of ganglioside GD(1b) are essential for the organization of lipid rafts in RBL-2H3 mast cells. *Exp. Cell Res.* 2008 Aug 1;314(13):2515-28.

[140] Simons K, Ehehalt R. Cholesterol, lipid rafts, and disease. *J. Clin. Invest.* 2002 Sep;110(5):597-603.

[141] Yamashita T, Yamaguchi T, Murakami K, Nagasawa S. Detergent-resistant membrane domains are required for mast cell activation but dispensable for tyrosine phosphorylation upon aggregation of the high affinity receptor for IgE. *Journal of biochemistry*. 2001 Jun;129(6):861-8.
[142] Kato N, Nakanishi M, Hirashima N. Cholesterol depletion inhibits store-operated calcium currents and exocytotic membrane fusion in RBL-2H3 cells. *Biochemistry*. 2003 Oct 14;42(40):11808-14.
[143] Fifadara NH, Aye CC, Raghuwanshi SK, Richardson RM, Ono SJ. CCR1 expression and signal transduction by murine BMMC results in secretion of TNF-alpha, TGFbeta-1 and IL-6. *International immunology*. 2009 Aug;21(8):991-1001.
[144] Bickel PE, Scherer PE, Schnitzer JE, Oh P, Lisanti MP, Lodish HF. Flotillin and epidermal surface antigen define a new family of caveolae-associated integral membrane proteins. *The Journal of biological chemistry*. 1997 May 23;272(21):13793-802.
[145] Volonte D, Galbiati F, Li S, Nishiyama K, Okamoto T, Lisanti MP. Flotillins/cavatellins are differentially expressed in cells and tissues and form a hetero-oligomeric complex with caveolins in vivo. Characterization and epitope-mapping of a novel flotillin-1 monoclonal antibody probe. *The Journal of biological chemistry*. 1999 Apr 30;274(18):12702-9.
[146] Kolter T, Proia RL, Sandhoff K. Combinatorial ganglioside biosynthesis. *The Journal of biological chemistry*. 2002 Jul 19;277(29):25859-62.
[147] Basciano LK, Berenstein EH, Kmak L, Siraganian RP. Monoclonal antibodies that inhibit IgE binding. *The Journal of biological chemistry*. 1986 Sep 5;261(25):11823-31.
[148] Guo NH, Her GR, Reinhold VN, Brennan MJ, Siraganian RP, Ginsburg V. Monoclonal antibody AA4, which inhibits binding of IgE to high affinity receptors on rat basophilic leukemia cells, binds to novel alpha-galactosyl derivatives of ganglioside GD1b. *The Journal of biological chemistry*. 1989 Aug 5;264(22):13267-72.
[149] Oliver C, Sahara N, Kitani S, Robbins AR, Mertz LM, Siraganian RP. Binding of monoclonal antibody AA4 to gangliosides on rat basophilic leukemia cells produces changes similar to those seen with Fc epsilon receptor activation. *J. Cell Biol.* 1992 Feb;116(3):635-46.

[150] Jamur MC, Grodzki AC, Berenstein EH, Hamawy MM, Siraganian RP, Oliver C. Identification and characterization of undifferentiated mast cells in mouse bone marrow. *Blood.* 2005 Jun 1;105(11):4282-9.

[151] Swaim WD, Minoguchi K, Oliver C, Hamawy MM, Kihara H, Stephan V, et al. The anti-ganglioside monoclonal antibody AA4 induces protein tyrosine phosphorylations, but not degranulation, in rat basophilic leukemia cells. *The Journal of biological chemistry.* 1994 Jul 29; 269(30):19466-73.

[152] Stracke ML, Basciano LK, Siraganian RP. Binding properties and histamine release in variants of rat basophilic leukemia cells with changes in the IgE receptor. *Immunology letters.* 1987 Apr;14(4):287-92.

[153] Stephan V, Guo N, Ginsburg V, Siraganian RP. Immunoprecipitation of membrane proteins from rat basophilic leukemia cells by the antiganglioside monoclonal antibody AA4. *J. Immunol.* 1991 Jun 15;146(12):4271-7.

[154] Minoguchi K, Swaim WD, Berenstein EH, Siraganian RP. Src family tyrosine kinase p53/56lyn, a serine kinase and Fc epsilon RI associate with alpha-galactosyl derivatives of ganglioside GD1b in rat basophilic leukemia RBL-2H3 cells. *The Journal of biological chemistry.* 1994 Feb 18;269(7):5249-54.

[155] Prieschl EE, Baumruker T. Sphingolipids: second messengers, mediators and raft constituents in signaling. *Immunology today.* 2000 Nov;21(11):555-60.

[156] Baumruker T, Prieschl EE. Sphingolipids and the regulation of the immune response. *Seminars in immunology.* 2002 Feb;14(1):57-63.

[157] Urtz N, Olivera A, Bofill-Cardona E, Csonga R, Billich A, Mechtcheriakova D, et al. Early activation of sphingosine kinase in mast cells and recruitment to FcepsilonRI are mediated by its interaction with Lyn kinase. *Molecular and cellular biology.* 2004 Oct;24(19):8765-77.

[158] Furne C, Corset V, Herincs Z, Cahuzac N, Hueber AO, Mehlen P. The dependence receptor DCC requires lipid raft localization for cell death signaling. *Proceedings of the National Academy of Sciences of the United States of America.* 2006 Mar 14;103(11):4128-33.

[159] Koenig A, Russell JQ, Rodgers WA, Budd RC. Spatial differences in active caspase-8 defines its role in T-cell activation versus cell death. *Cell death and differentiation.* 2008 Nov;15(11):1701-11.

[160] Galli SJ. New insights into "the riddle of the mast cells": microenvironmental regulation of mast cell development and phenotypic heterogeneity. *Laboratory investigation; a journal of technical methods and pathology*. 1990 Jan;62(1):5-33.

[161] Yong LC. The mast cell: origin, morphology, distribution, and function. *Exp. Toxicol. Pathol.* 1997 Dec;49(6):409-24.

[162] Passante E, Frankish N. The RBL-2H3 cell line: its provenance and suitability as a model for the mast cell. *Inflamm. Res.* 2009 Nov; 58(11):737-45.

[163] Silveira e Souza AM, Trindade ES, Jamur MC, Oliver C. Gangliosides are important for the preservation of the structure and organization of RBL-2H3 mast cells. *J. Histochem. Cytochem.* 2010 Jan;58(1):83-93.

[164] Levitan I, Gooch KJ. Lipid rafts in membrane-cytoskeleton interactions and control of cellular biomechanics: actions of oxLDL. *Antioxidants & redox signaling*. 2007 Sep;9(9):1519-34.

[165] Bunnell SC, Kapoor V, Trible RP, Zhang W, Samelson LE. Dynamic actin polymerization drives T cell receptor-induced spreading: a role for the signal transduction adaptor LAT. *Immunity*. 2001 Mar;14(3):315-29.

[166] Barda-Saad M, Braiman A, Titerence R, Bunnell SC, Barr VA, Samelson LE. Dynamic molecular interactions linking the T cell antigen receptor to the actin cytoskeleton. *Nature immunology*. 2005 Jan;6(1):80-9.

[167] Gomez TS, Hamann MJ, McCarney S, Savoy DN, Lubking CM, Heldebrant MP, et al. Dynamin 2 regulates T cell activation by controlling actin polymerization at the immunological synapse. *Nature immunology*. 2005 Mar;6(3):261-70.

[168] Lenne PF, Wawrezinieck L, Conchonaud F, Wurtz O, Boned A, Guo XJ, et al. Dynamic molecular confinement in the plasma membrane by microdomains and the cytoskeleton meshwork. *Embo J.* 2006 Jul 26;25(14):3245-56.

[169] Chichili GR, Rodgers W. Clustering of membrane raft proteins by the actin cytoskeleton. *The Journal of biological chemistry*. 2007 Dec 14;282(50):36682-91.

[170] Chichili GR, Rodgers W. Cytoskeleton-membrane interactions in membrane raft structure. *Cell Mol. Life Sci.* 2009 Jul;66(14):2319-28.

[171] Yin HL, Janmey PA. Phosphoinositide regulation of the actin cytoskeleton. *Annu. Rev. Physiol.* 2003;65:761-89.

[172] Czech MP. Lipid rafts and insulin action. *Nature*. 2000 Sep 14; 407(6801):147-8.

[173] Harlan JE, Hajduk PJ, Yoon HS, Fesik SW. Pleckstrin homology domains bind to phosphatidylinositol-4,5-bisphosphate. *Nature*. 1994 Sep 8;371(6493):168-70.

[174] Varnai P, Balla T. Visualization of phosphoinositides that bind pleckstrin homology domains: calcium- and agonist-induced dynamic changes and relationship to myo- [3H]inositol-labeled phosphoinositide pools. *J. Cell Biol.* 1998 Oct 19;143(2):501-10.

[175] Lemon G, Gibson WG, Bennett MR. Metabotropic receptor activation, desensitization and sequestration-I: modelling calcium and inositol 1,4,5-trisphosphate dynamics following receptor activation. *J. Theor. Biol.* 2003 Jul 7;223(1):93-111.

[176] Funatsu N, Kumanogoh H, Sokawa Y, Maekawa S. Identification of gelsolin as an actin regulatory component in a triton insoluble low density fraction (raft) of newborn bovine brain. *Neuroscience research*. 2000 Apr;36(4):311-7.

[177] Ilschner S, Brandt R. The transition of microglia to a ramified phenotype is associated with the formation of stable acetylated and detyrosinated microtubules. *Glia*. 1996 Oct;18(2):129-40.

[178] Conde C, Caceres A. Microtubule assembly, organization and dynamics in axons and dendrites. *Nature reviews*. 2009 May;10(5):319-32.

[179] Tsan MF, Berlin RD. Effect of phagocytosis on membrane transport of nonelectrolytes. *J. Exp. Med.* 1971 Oct 1;134(4):1016-35.

[180] Edelman GM, Yahara I, Wang JL. Receptor mobility and receptor-cytoplasmic interactions in lymphocytes. *Proceedings of the National Academy of Sciences of the United States of America*. 1973 May;70(5):1442-6.

[181] Oliver JM, Ukena TE, Berlin RD. Effects of phagocytosis and colchicine on the distribution of lectin-binding sites on cell surfaces. *Proc. Natl. Acad. Sci. U. S. A.* 1974 Feb;71(2):394-8.

[182] Ceresa BP, Schmid SL. Regulation of signal transduction by endocytosis. *Current opinion in cell biology*. 2000 Apr;12(2):204-10.

[183] Brodsky FM, Chen CY, Knuehl C, Towler MC, Wakeham DE. Biological basket weaving: formation and function of clathrin-coated vesicles. *Annu. Rev. Cell Dev. Biol.* 2001;17:517-68.

[184] Mousavi SA, Malerod L, Berg T, Kjeken R. Clathrin-dependent endocytosis. *Biochem. J.* 2004 Jan 1;377(Pt 1):1-16.

[185] Sorkin A. Cargo recognition during clathrin-mediated endocytosis: a team effort. *Current opinion in cell biology*. 2004 Aug;16(4):392-9.

[186] Le Roy C, Wrana JL. Clathrin- and non-clathrin-mediated endocytic regulation of cell signalling. *Nat. Rev. Mol. Cell Biol.* 2005 Feb;6(2): 112-26.

[187] Ewers H, Helenius A. Lipid-mediated endocytosis. *Cold Spring Harb. Perspect. Biol.* 2011 Aug;3(8):a004721.

[188] Mayor S, Pagano RE. Pathways of clathrin-independent endocytosis. *Nat. Rev. Mol. Cell Biol.* 2007 Aug;8(8):603-12.

[189] Molfetta R, Gasparrini F, Santoni A, Paolini R. Ubiquitination and endocytosis of the high affinity receptor for IgE. *Mol. Immunol.* 2010 Sep;47(15):2427-34.

[190] Stump RF, Pfeiffer JR, Seagrave J, Oliver JM. Mapping gold-labeled IgE receptors on mast cells by scanning electron microscopy: receptor distributions revealed by silver enhancement, backscattered electron imaging, and digital image analysis. *J. Histochem. Cytochem.* 1988 May;36(5):493-502.

[191] Mao SY, Pfeiffer JR, Oliver JM, Metzger H. Effects of subunit mutation on the localization to coated pits and internalization of cross-linked IgE-receptor complexes. *J. Immunol.* 1993 Sep 1;151(5):2760-74.

[192] Bonifacino JS, Traub LM. Signals for sorting of transmembrane proteins to endosomes and lysosomes. *Annu. Rev. Biochem.* 2003;72:395-447.

[193] Mazucato VM, Silveira ESAM, Nicoletti LM, Jamur MC, Oliver C. GD1b-derived gangliosides modulate FcepsilonRI endocytosis in mast cells. *J. Histochem. Cytochem.* 2011 Apr;59(4):428-40.

[194] Di Fiore PP, De Camilli P. Endocytosis and signaling. an inseparable partnership. *Cell.* 2001 Jul 13;106(1):1-4.

[195] Lafont F, Simons K. Raft-partitioning of the ubiquitin ligases Cbl and Nedd4 upon IgE-triggered cell signaling. *Proc. Natl. Acad. Sci. U. S. A.* 2001 Mar 13;98(6):3180-4.

[196] Harvey KF, Kumar S. Nedd4-like proteins: an emerging family of ubiquitin-protein ligases implicated in diverse cellular functions. *Trends Cell Biol.* 1999 May;9(5):166-9.

[197] Freemont PS. RING for destruction? *Curr. Biol.* 2000 Jan 27;10(2):R84-7.

[198] Ota Y, Beitz LO, Scharenberg AM, Donovan JA, Kinet JP, Samelson LE. Characterization of Cbl tyrosine phosphorylation and a Cbl-Syk complex in RBL-2H3 cells. *J. Exp. Med.* 1996 Nov 1;184(5):1713-23.

[199] Ota Y, Samelson LE. The product of the proto-oncogene c-cbl: a negative regulator of the Syk tyrosine kinase. *Science.* 1997 Apr 18;276(5311):418-20.

[200] Molfetta R, Gasparrini F, Peruzzi G, Vian L, Piccoli M, Frati L, et al. Lipid raft-dependent FcepsilonRI ubiquitination regulates receptor endocytosis through the action of ubiquitin binding adaptors. *PLoS One.* 2009;4(5):e5604.

[201] Jamur MC, Grodzki AC, Moreno AN, de Mello Lde F, Pastor MV, Berenstein EH, et al. Identification and isolation of rat bone marrow-derived mast cells using the mast cell-specific monoclonal antibody AA4. *J. Histochem. Cytochem.* 2001 Feb;49(2):219-28.

[202] Jamur MC, Grodzki AC, Moreno AN, Swaim WD, Siraganian RP, Oliver C. Immunomagnetic isolation of rat bone marrow-derived and peritoneal mast cells. *J. Histochem. Cytochem.* 1997 Dec;45(12):1715-22.

[203] Faraco CD, Vugman I, Siraganian RP, Jamur MC, Oliver C. Immunocytochemical identification of immature rat peritoneal mast cells using a monoclonal antibody specific for rat mast cells. *Acta histochemica.* 1997 Mar;99(1):23-7.

Reviewed by Devandir Antonio de Souza Junior, PhD, Department of Cell and Molecular Biology and Pathogenic Bioagents, School of Medicine of Ribeirão Preto, University of São Paulo, Ribeirão Preto, SP, Brazil.

In: Lipid Rafts
Editor: Dan Sillence

ISBN: 978-1-62948-961-2

Chapter VII

The Role of Lipid Rafts in Neurodegenerative Diseases

Clive Bate[*]
Department of Pathology and Pathogen Biology, Royal Veterinary College, Hawkshead Lane, North Mymms, Herts, UK

Abstract

Lipid rafts are specific membrane micro-domains that are enriched in cholesterol, sphingolipids and a distinct set of proteins, including those attached to cell membranes by glycosylphosphatidyinositol (GPI) anchors. They are involved in cell function through the regulation of protein trafficking, endocytosis, exocytosis and cell signalling. Lipid rafts are sensitive to cholesterol depletion and in neurons the presence of cholesterol is critical in the formation of synapses and the regulation of synaptic transmission. Many neurodegenerative diseases have been shown to be associated with altered cholesterol homeostasis and synapse dysfunction. The dementia associated with Alzheimer's, Parkinson's and prion diseases is caused by specific proteins/peptides that disrupt synapse function. In this chapter the interactions between these disease-associated proteins/peptides and lipid rafts, and the emerging role of lipid rafts in the molecular mechanisms that lead to aberrant cell signalling, synapse degeneration and consequently the clinical symptoms of these diseases is discussed.

[*] Corresponding author: email: cbate@rvc.ac.uk.

Many disease-associated proteins/peptides accumulate within lipid rafts and trigger the formation of signalling platforms. The cellular prion protein (PrP^C) that is targeted to lipid rafts by its GPI anchor is implicated in the pathogenesis of both prion and Alzheimer's diseases. Here I discuss the hypothesis that PrP^C acts as a scaffold protein that triggers the formation of signalling platforms that incorporate and activate cytoplasmic phospholipase A_2 ($cPLA_2$). The activation of $cPLA_2$ results in a cascade of mediators including platelet-activating factor (PAF) and prostaglandins (PG)s. While these factors regulate normal neurotransmission, the overproduction of PAF and PGE_2 triggers synapse degeneration. More specifically, the aggregation of PrP molecules, following self-aggregation of the disease-associated isoform (PrP^{Sc}) or cross-linkage of PrP^C by amyloid-β_{42} oligomers (that are thought to constitute the major neurotoxin in Alzheimer's disease) triggers the formation of raft/complexes and activation of $cPLA_2$. A detailed understanding of the molecular mechanisms by which lipid rafts coalesce into these signalling platforms and subsequently disperse is vital to the discovery of new drugs.

In the second part of this chapter I discuss several compounds that might modify the formation and/or dissociation of lipid rafts and so may be helpful in ameliorating the effects of prions or Aβ on synapses. These compounds include both prescription drugs and nutraceuticals raising the possibility that modifications of diet may alter the progression of some neurodegenerative diseases.

Lipid Rafts – Difficult Definitions?

For the purposes of this chapter it is helpful to imagine lipid rafts as small patches of membranes that "float" within the cell membrane. Under specific stimuli individual lipid rafts coalesce into larger platforms that regulate cell signalling. Such rafts typically constitute only a small percentage of the cell membrane. A commonly used method to define lipid rafts is by their insolubility in cold non-ionic detergents such as Triton X-100, Lubrol or Brij [1,2] giving rise to alternative terminologies such detergent-resistant membranes (DRMs) or detergent-insoluble glycodomains (DIGs). Different mixtures and concentrations of detergents can be used to isolate lipid rafts under more physiological conditions [3] and detergent-free methods have also been developed [4-6]. Caution should be used in the interpretation of such data as the use of different methods results in different protein compositions [2]. Thus, biochemical results are usually combined with visualisation techniques as recently reviewed by Simons and Gerl [7] using defined "raft-resident" molecules such as caveolins, cholera toxin subunit B (CTxB) for ganglioside GM-1, or the fluorescent cholesterol analogue dehydroergosterol [8]. In 2006 a working definition of lipid rafts was developed at the Keystone symposium on lipid rafts and cellular function stating that "membrane rafts are small (10-200 nm), heterogeneous, highly dynamic, sterol- and sphingolipid-enriched domains that compartmentalize cellular processes. Small rafts can sometimes be stabilised to form larger platforms through protein-protein and protein-lipid interactions" [9].

Lipid Rafts – Key Components

Although cholesterol is distributed throughout cell membranes, its concentration is up to 5 fold greater in lipid rafts [10]. Cholesterol can modify the physiochemical properties and molecular organisation of membranes so that any disturbances in its metabolism can lead to altered membrane organisation and cellular dysfunction. Lipid rafts are also highly enriched in sphingolipids, gangliosides [11,12] and phospholipids containing saturated fatty acids [13]. One key property of lipid rafts is that they include or exclude specific proteins. The repertoire of raft-resident proteins includes proteins anchored to cell membranes by glycosylphosphatidylinositol (GPI) anchors [14-17], palmitoylated proteins [18] and a number of structural proteins including caveolins and flotillins [19]. GPI anchors and sphingolipids are found within lipid rafts as they contain saturated fatty acids that allow tight molecular packing and increase the solubilisation of cholesterol [20-22], and the glycan components protect cholesterol from water [23]. Both GPI-anchored proteins and sphingolipids are thought to help maintain the composition and consequently the function of rafts [24].

Lipid rafts are heterogenous; different rafts contain different proteins and glycolipids [10,25] and consequently have different functions. Some lipid rafts contain cellular receptors, such as the folate and p75 neurotrophin receptors [26,27] and receptors for neurotransmitters including acetylcholine [28] and gamma-aminobutyric acid [29]. They also act as membrane platforms that concentrate molecules for cell signalling [30] and contain Src family tyrosine kinases [31,32], adenylyl cyclase [33], the trimeric G-proteins [34], phosphatidylinositol bisphosphate [35], Ras proteins [36], cytoplasmic phospholipase A_2 ($cPLA_2$) [37] platelet-activating factor (PAF) [38] and redox signalling enzymes [39].

Lipid rafts are thought to exist as nanoscale micro-domains that range from 5 to 200 nm in diameter and contain only a few proteins [10,40]. Such membranes are "poised" to form larger scale rafts or signalling platforms [41] and they can coalesce under specific stimuli to form larger rafts with complex interactions between constituent proteins, lipids and glycolipids [42-44]. The formation of such rafts/platforms causes co-patching of some cellular components and the exclusion of others [45]. Raft formation is associated with the oligomerization of proteins [46]. For example, the oligomerization of raft proteins by antibodies or polyvalent toxins [47] can combine low affinity lipid-interacting molecules into a stable, functional lipid raft. The crosslinking of membrane components leads to large-scale membrane reorganization involving numerous other membrane components [41,48]. Thus rafts induced by the clustering of a small number of proteins or lipids can rapidly reorganize many other membrane proteins. The clustering of separate rafts exposes proteins to new membrane environments whereby even small changes in partitioning into lipid rafts can amplify responses and initiate signalling cascades. Notably the organisation of rafts in the outer membrane leaflet is reflected in changes in the inner/cytoplasmic leaflet [49,50] including the recruitment of molecules such as the Ras family of GTPases and tyrosine kinases [51].

Given that the formation and function of lipid rafts depends upon the amount of cholesterol within cell membranes [10,43,52], it follows that fluctuations in the cholesterol content of cells alters the properties of lipid rafts. Consequently lipid rafts may be susceptible to cholesterol depletion that results in the release of raft constituents into the plasma membrane. The inhibition of cell signalling following cholesterol depletion is a defining characteristic of raft-associated signalling pathways.

Alzheimer's Disease

Alzheimer's disease (AD), the most common form of dementia, is a genetically heterogeneous, complex, multi-faceted neurodegenerative disease characterized by progressive cognitive decline resulting from synapse damage [53,54]. The amyloid hypothesis of AD pathogenesis maintains that the main event leading to dementia in AD is the production of amyloid-β (Aβ) peptides derived from proteolytic cleavage of the amyloid precursor protein (APP) by β and γ secretases [55]. The accumulation of Aβ leads to the subsequent disruption of neuronal processes, abnormal phosphorylation of tau, synapse damage and ultimately the death of neurons [56]. Aβ peptides added to cultured neurons can mimic some of the events that characterise neurodegeneration in AD including tau hyperphosphorylation [57] and synapse damage [58,59].

Although brains from Alzheimer's patients contain a toxic soup of Aβ peptides [60], the 42 amino acid fragment ($A\beta_{42}$) is widely regarded as the major neurotoxin [61]. $A\beta_{42}$ fragments self-aggregate and consequently exist in multiple forms ranging from monomers to oligomers, fibrils and plaques. Different conformational forms of $A\beta_{42}$ exist and it is recognised that there are specific pools of $A\beta_{42}$ that are disease-relevant, while other conformations/pools are less toxic or biologically inert. The soluble $A\beta_{42}$ oligomers are thought to be responsible for synapse damage and memory defects [62-65].

Cholesterol and Lipid Rafts in Alzheimer's Disease

Numerous epidemiological, biochemical and genetic studies suggest that the concentration and distribution of cholesterol regulates the progression of AD [66-72]. Cholesterol is required for the formation of lipid rafts [52] and raft-dependent processes are considered to be critical factors in the pathogenesis of AD. The role of cholesterol and lipid rafts in the production of $A\beta_{42}$ has received most attention and has been reviewed elsewhere [73,74]. Other studies have demonstrated that once formed $A\beta_{42}$ accumulates within lipid rafts [75-78] and that Aβ disturbs cellular cholesterol homeostasis and can modify cell membrane biochemistry [79-82]. Hypercholesterolaemia accelerated amyloid pathology in a transgenic mouse model [83] and drugs that inhibit cholesterol synthesis slow the onset and progression of AD [69,84]. Drug-induced cholesterol depletion *in vitro* protected neurons against $A\beta_{42}$-induced neurodegeneration [85,86] and reduced $A\beta_{42}$ formation [87], suggesting that both $A\beta_{42}$ formation and $A\beta_{42}$-induced neurodegeneration are generated from within lipid rafts.

The amounts of cholesterol within cell membranes is controlled through the biosynthesis, uptake and efflux to and from the extracellular medium and by esterification of cholesterol in the endoplasmic reticulum (ER) [88]. A dynamic equilibrium between the pools of cholesterol and cholesterol esters is controlled by acyl-coenzyme A:cholesterol acyltransferase (ACAT), an ER-resident enzyme that catalyses the formation of cholesterol esters from cholesterol and fatty acids [89] and by cholesterol ester hydrolases (CEH) which release cholesterol from cholesterol esters. Observations that ACAT inhibitors reduced the levels of Aβ in a transgenic AD mouse model raised interest in such drugs as possible treatments for Alzheimer's disease [90-92].

Prion Diseases

The transmissible spongiform encephalopathies, more commonly known as "prion diseases", include scrapie and bovine spongiform encephalopathy (BSE), also known as "Mad Cow disease" have gained attention for two reasons. The first is that they represent a remarkable form of infectious agent that appears to be a self-replicating protein [93]. Secondly, in the 1990's the observation that new variant Creutzfeldt-Jakob disease (vCJD), a human prion disease, was associated with BSE suggested that these prions crossed the species barrier, raising fears of an epidemic. Fortunately those early predictions were mostly remarkable for their inaccuracy and so far fewer than 200 vCJD cases have been confirmed.

Prion diseases occur when the normal cellular prion protein (PrP^{C}) is misfolded into a disease-associated isoform (PrP^{Sc}) [93,94]. The newly formed PrP^{Sc} acts as a catalyst for the conversion of PrP^{C} to PrP^{Sc}, resulting in the gradual accumulation of PrP^{Sc}, within the brain which is thought to cause the synapse damage and the symptoms of the prion diseases. In cultured neurons synapse damage is triggered by small soluble PrP^{Sc} oligomers [95] and by a peptide containing amino acids 82 to 146 of the PrP protein (PrP82-146), found in the brains of patients with the human prion disease, Gerstmann-Sträussler-Scheinker disease [96].

Cholesterol and Rafts in Prion Disease

Prion infection is associated with changes in cholesterol levels, both in the brain [97] and in neuronal cell lines [98]. The aggregation of PrP^{Sc} affects the composition of cell membranes as the clustering of GPI anchors attached to PrP^{Sc} increases the solubility of cholesterol in membranes. This is a rare example of an infective agent increasing cholesterol concentrations within cell membranes, and it raises the possibility that the pathology of prion diseases may be triggered by disturbances in membrane cholesterol.

The presence of GPI anchors targets PrP^{C} and PrP^{Sc} to lipid rafts [99-101]. The importance of lipid rafts in prion diseases is based upon studies showing that treatment with cholesterol synthesis inhibitors reduced the formation of PrP^{Sc} [102-104]. Prion infection is associated not only with increased amounts of cholesterol in the cell membrane, but also with reduced amounts of cholesterol esters. These changes in prion-infected cells may be caused by direct inhibition of ACAT, or by sequestration of cholesterol in microenvironments that avoid ACAT. Such changes in cell membranes would reduce both endocytosis and the recycling of cholesterol through the ER where it is exposed to ACAT, consistent with the reduced cholesterol ester production of infected cells.

Role of Cholesterol and Rafts in Other Neurodegenerative Diseases

Disruption of cholesterol metabolism within cells has profound effects on cell function, interfering with cell signalling and hence causing pathology. Dysregulation of cholesterol levels within the brain can affect the pathogenesis of Parkinson's diseases, dementia with Lewy bodies, α-synucleinopathies, epilepsy and multiple sclerosis [67,71,105]. However while treatment with cholesterol synthesis inhibitors (statins) has been shown to be beneficial in experimental models of Parkinson's disease [106-108], their effects on patients with Parkinson's disease remain controversial [109].

Dementia in a Culture Dish? - Measuring Synapse Damage in Neuronal Cultures

Some of the pathological characteristics of neurodegenerative diseases can be replicated *in vitro* by incubating cultured neurons with disease-associated peptides. For example, as dementia in AD is associated with the loss of synaptic proteins [110-118] synapse damage can be examined by measuring the amounts of synaptic proteins in cultured neurons [119]. The addition of $A\beta_{42}$ oligomers reduced the amounts of synaptophysin, synapsin-1 and vesicle-associate membrane protein (VAMP)-1 in cultured neurons [120] providing a reliable tissue culture model of the synapse damage observed in AD. Effects were observed at concentrations of $A\beta_{42}$ comparable to those present in the brains of Alzheimer's patients [121-124]. Similar dose-dependent decreases in synaptic protein concentrations were observed in neurons incubated with PrP^{Sc} [95]. Prion and $A\beta_{42}$-induced synapse dysfunction can also be measured by examining synaptic vesicle recycling in terms of the uptake of fluorescent dyes. Notably the addition of picomolar $A\beta_{42}$ oligomers blocked synaptic vesicle recycling in cultured neurons [120].

Studies with cultured neurons were complemented by experiments in which the effects of disease-related peptides on isolated synapses (synaptosomes) were examined. The effects of PrP^{Sc} or $A\beta_{42}$ on cholesterol, lipid raft composition and cell signalling pathways in synaptosomes were studied.

Lipid Rafts in the Formation of Signalling Platforms

It is thought that multiple individual lipid rafts coalesce to form larger platforms capable of cell activation [10,41]. The coalescence of individual lipid rafts can orchestrate cell signalling by bringing together receptors and signalling proteins, signalling molecules that act cooperatively or enzymes with specific substrates. Some examples include the formation of CD14:MD2:TLR4 complexes by myeloid cells in response to lipopolysaccharide [125] and complexes that mediate IgE, T and B cell receptor signalling [30,126-128]. A small change in the clustering of specific membrane proteins can lead to large-scale membrane reorganization that involves numerous other membrane components [47]. While many components of these complexes, such as CD14, are resident in rafts, other key signalling molecules such TLR4, some G proteins and Src family tyrosine kinases are recruited to rafts following engagement of receptors [51,125,129,130]. The formation of raft platforms results in increased receptor density within a small membrane area, rafts limit the lateral diffusion of receptors and consequently stabilise interactions between ligand and receptors. Lipid rafts are present in both the outer surface and inner (cytoplasmic) leaflets of the cell membrane. As a key to signal transduction the reorganisation of lipid rafts in the outer leaflet in response to external stimuli causes reorganisation of the inner, cytoplasmic leaflet membrane and the sorting of intracellular signalling molecules [49-51,130]. The observation that many "raft-associated proteins" are found in non-raft membranes in quiescent cells suggests that the dissociation of raft platforms, resulting in the cessation of cell signalling, is a mechanism that limits cell activation. Moreover conditions in which cell signalling platforms fail to dissociate may lead to sustained activation of signalling pathways, cell disruption and disease.

Cholesterol is Required for Normal Synapse Function

Cholesterol is supplied to neurons by glial cells; it is highly enriched in synaptic membranes where it is required for synapse formation [131-135]. The fine tuning of cholesterol dynamics is thought to underlie synapse plasticity and hence memory as the amounts of cholesterol in synaptic membranes affects synaptic vesicle recycling [136,137] and the sensitivity of neurotransmitter receptors [29,138].

Role of $cPLA_2$ in Normal Synapse Function

Synapses are highly enriched in $cPLA_2$, an enzyme that regulates neurotransmission and is involved in the acquisition of short term memory [139-142]. It is thought that $cPLA_2$ has two major functions at the synapse. The first is to regulate synapse structure; activated $cPLA_2$ results in the formation of lyso-phospholipids, which control membrane curvature and hence endo and exocytosis [143,144]. As the accumulation of large amounts of lyso-phospholipids results in cell lysis the levels of lyso-phospholipids in cell membranes are controlled by lyso-phospholipid acyl transferases (LPATs), enzymes that reacylate lyso-phospholipids. LPATs such as the endophilins are also concentrated within synapses and regulate synaptic vesicle recycling [145-147].

Activated $cPLA_2$ also causes the release of arachidonic and docsahexaenoic acids from the sn-2 position of specific phospholipids. Arachidonic acids are metabolised to prostaglandins (PG)s, including PGE_2 which regulates synaptic plasticity [148-150], leucotrienes and lipoxins [151]. Docosahexaenoic acid can be converted to neuroprotective metabolites [152-154]. The activation of $cPLA_2$ is also the first step in the production of platelet-activating factor (PAF) [151,155], that is produced in response to specific stimuli [156,157]. PAF receptors are concentrated in synapses [158] and PAF controls synaptic plasticity and regulates memory formation [159-162].

The Role of $cPLA_2$ in Neurodegenerative Diseases

The aberrant activation of $cPLA_2$ has been implicated in the pathogenesis of AD based on reports that;

i) $A\beta_{42}$ activates $cPLA_2$ [163-165]
ii) That inhibition of $cPLA_2$ reduced the $A\beta_{42}$-induced synapse damage *in vitro* [120] and *in vivo* [166].
iii) That inhibition of $cPLA_2$ ameliorated cognitive defects in a mouse model of AD [167].

Similarly, a critical role of $cPLA_2$ in prion diseases is suggested by;

i) Significant correlations between the amounts of PrP^{Sc}, cholesterol and activated $cPLA_2$ in cell lines.
ii) Studies showing that activated $cPLA_2$ is targeted to PrP^{Sc}-containing lipid rafts [98]
iii) That inhibition of $cPLA_2$ reduced both PrP^{Sc} formation [168] and the neurotoxicity of PrP^{Sc} [169].

However, *in vivo* studies are hampered by the pharmacokinetic properties of common $cPLA_2$ inhibitors as they do not readily penetrate the blood-brain-barrier [170,171].

The hypothesis that aberrant activation of $cPLA_2$ triggers neurodegeneration is supported by analyses of lipid mediators such as PAF and PGs that are released after activation of $cPLA_2$. PGE_2 is a marker of neurodegeneration in prion diseases; it is increased in brain areas showing PrP^{Sc} accumulation in experimental scrapie [172] and is raised in the cerebrospinal fluid of patients with Creutzfeldt-Jakob disease [173,174]. Similarly, concentrations of PAF and PGE_2 are raised in the cerebrospinal fluid of patients with AD [175,176]. *In vitro* studies showed that physiologically relevant concentrations of PAF and PGE_2 caused synapse damage [120,177]. Consequently there is interest in the development of PG and PAF antagonists for use in neurological disease [178]. Notably, in a murine AD model, transgenic mice that lacked the EP2 receptor for PGE_2 had reduced Aβ [179].

The $cPLA_2$ cascade has been implicated in other neurodegenerative diseases including bipolar disorders [180], epilepsy [181,182] and schizophrenia [183]. The neuropathology induced in the MPTP-model of Parkinson's disease can be blocked by $cPLA_2$ inhibitors [184,185]. Dementia is a common symptom associated with HIV infections [186,187] because viral-encoded proteins cause synapse dysfunction [188-190]. HIV-1-transgenic rats show signs of activated $cPLA_2$ [191] and PAF antagonists are being developed to treat HIV-related dementia [192-194].

Activation of $cPLA_2$ is Cholesterol-Sensitive

The enzyme $cPLA_2$ can be activated in both a cell and stimulus specific manner. Its activation is accompanied by its translocation from the cytoplasm to specific intracellular membranes utilizing an N-terminal, calcium-dependent lipid binding domain [195-198]. This translocation of $cPLA_2$ is dependent upon cholesterol-sensitive membranes [37,98,199]. More specifically, the $A\beta_{42}$-induced activation of $cPLA_2$ in synapses involves its translocation from the cytoplasm to lipid rafts [200] and in prion-infected cells $cPLA_2$ is found within PrP^{Sc}-containing rafts [98]. Reports that cholesterol depletion reduces the $A\beta_{42}$-induced activation of $cPLA_2$ [85] and triggers the dispersal of $cPLA_2$ from PrP^{Sc} rafts [98] confirms the critical role of cholesterol in prion and Aβ-induced activation of $cPLA_2$.

The Role of PrP^C in Alzheimer's Disease

The high potency of $A\beta_{42}$ oligomers and the accumulation of Aβ in synapses [201-203] suggest that $A\beta_{42}$-induced synapse damage is a receptor-mediated process. The identification of specific Aβ receptors is therefore key to understanding the pathogenesis of AD. Several candidates have been proposed, including NMDA and mGlu5R receptors [58,204], the amyloid precursor protein [205] and the receptor for advanced glycation end products (RAGE) [206]. However, the role of each of these proteins in the biological activity of Aβ remains unclear. More recently PrP^C was identified as a receptor that mediated $A\beta_{42}$-induced inhibition of synaptic plasticity and impaired memory in a model of AD [207]. PrP^C is a recycling protein that is found in lipid rafts, traffics between the cell surface and intracellular compartments [208] and is highly expressed at synapses [209,210].

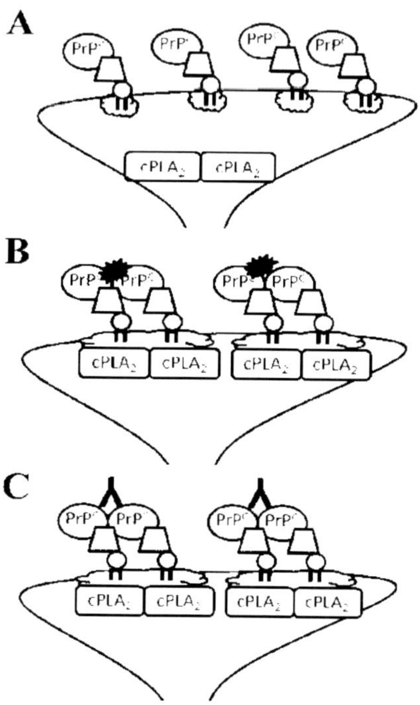

Figure 1. PrP^C mediates the activation of $cPLA_2$ and synapse damage induced by Aβ oligomers. Cartoon showing the proposed relationship between PrP^C, $cPLA_2$, lipid rafts ( ) and Aβ oligomers ( ). (A) In the absence of Aβ oligomers, $cPLA_2$ and PrP^C and are not associated, there is no activation of $cPLA_2$ or synapse damage. (B) Aβ oligomers cross-link PrP^C resulting in a complex containing Aβ, PrP^C and $cPLA_2$ leading to the activation of $cPLA_2$ and synapse damage. (C) The addition of mAb 4F2 also cross-links PrP^C resulting in the association between $cPLA_2$ and PrP^C, activation of $cPLA_2$ and synapse damage.

PrP^C is thought to act as a scaffold protein that organises signalling platforms [211]. Consistent with the hypothesis that cross-linkage of lipid raft components triggers phase separation and cell signalling [47], cross-linkage of

PrP^C by Aβ oligomers induced the activation of $cPLA_2$ and synapse damage in cultured neurons [201]. Observations that aggregated PrP^C causes synaptic abnormalities [212] and that cross-linkage of PrP^C by mAbs leads to the formation of lipid rafts, cell signalling and neurodegeneration [213,214] raise the possibility that it is the cross-linkage of PrP^C that leads to synapse damage. This hypothesis is supported by reports that Aβ oligomers (with the potential to cross-link PrP^C) cause synapse dysfunction [62,215] whereas Aβ monomers are generally considered non-toxic [59,216]. Similarly, cross-linkage of PrP^C by mAb 4F2 activated $cPLA_2$, whereas monovalent Fab fragments of mAb 4F2 had no significant affect (Figure 1).

In synaptosomes, Aβ oligomer-induced cross-linkage of PrP^C caused an increase in the amount of cholesterol, activation of $cPLA_2$ and its consequent migration to PrP^C-containing rafts [201]. Pharmacological inhibition of cholesterol ester hydrolysis not only blocked the Aβ-induced increase in the amount of cholesterol, but also reduced Aβ-induced activation of $cPLA_2$, indicating that Aβ-induced hydrolysis of cholesterol esters in synapses provides cholesterol to stabilise the Aβ:$cPLA_2$:PrP^C complex. This hypothesis is consistent with the hypothesis that critical concentrations of $Aβ_{42}$ cross-link PrP^C and trigger the formation of raft platforms that capture $cPLA_2$, causing sustained activation of $cPLA_2$ and consequent synapse damage.

Compounds that Act Directly upon the $cpla_2$ Cascade

So far few inhibitors of $cPLA_2$ have been developed to treat AD, possibly because there is a lack of suitable compounds that penetrate the blood-brain barrier (BBB). There is increasing interest in valproic acid, (2-propylpentanoic acid) a short branched-chain fatty acid used for the treatment of epilepsy that readily crosses the BBB. Its therapeutic role has expanded to include migraine and bipolar disorder prophylaxis, and recent research suggests a potential new roles in AD and in dementia associated with Parkinson's disease and HIV infections [217-219]. Reports that valproic acid affects arachidonic acid release [220] suggest that it affects the activation of $cPLA_2$. Preliminary studies demonstrated that valproic acid suppressed $Aβ_{42}$ oligomer and prion-induced activation of $cPLA_2$ in synaptosomes and protected cultured neurons against $Aβ_{42}$ oligomer and prion-induced synapse damage (manuscript in preparation). Sequestration of signalling proteins away from their endogenous substrates is an emerging concept in cell regulation and valproic acid was also found to prevent $A\beta_{42}$-induced translocation of $cPLA_2$ to lipid rafts.

Activation of PLA_2 leads to the formation of leucotrienes, PGs and PAF. Concentrations of PGE_2 and PAF are raised in the brains of AD patients [175,176] and pico/nanomolar concentrations of PAF and PGE_2 cause synapse damage in cultured neurons [120,177,221]. Notably, synapse damage induced by either Aβ, prions or PAF was blocked by PGE_2 receptor antagonists [120]. It is noteworthy that PAF antagonists such as ginkgolide B (a component of the herbal remedy *ginkgo Biloba,* extracts of which are widely used to improve cognition in those with mild dementia/AD) [222-224], reduce Aβ-induced synapse damage in cultured neurons [221]. Similarly, some of the antiviral drugs that are used to treat HIV infections have PAF receptor antagonist activity and protect against HIV-associated dementia [192-194].

Pharmacological Manipulation of Lipid Rafts as a Novel Method of Neuroprotection

The identification of drugs that protect neurons against $A\beta_{42}$-induced synapse damage is a rational approach to reduce the cognitive decline of AD. For example, insulin reduced the binding of $A\beta_{42}$ to synapses in cultured neurons [225] and subsequent clinical trials indicate that insulin improves cognition in AD [226]. Conventional approaches have concentrated upon identifying compounds that have direct actions on target proteins, either by competitive or non-competitive inhibition of an active site. As cell signalling in lipid rafts is sensitive to the composition of the micro-environment, then compounds that affect lipid raft components may also affect raft-dependent signalling. The most common example of this type of inhibition is that by cholesterol synthesis inhibitors; by reducing cholesterol production they can affect the formation of lipid rafts and therefore signalling events that occur within them.

Cholesterol Synthesis Inhibitors

Numerous epidemiological, biochemical and genetic studies have demonstrated that cholesterol is a risk factor for the development of AD [66-

68]. Aβ disturbs cellular cholesterol homeostasis [79,80] and drugs that inhibit cholesterol synthesis slow the onset and progression of AD [69,84]. Observations that Aβ peptides are found within lipid rafts within the brain and in cultured neurons [76-78] and that cholesterol depletion protects neurons against Aβ-induced toxicity [85,86] suggest that Aβ oligomers trigger the events leading to neurodegeneration from within lipid rafts. Similarly, prions and toxic PrP peptides are targeted to lipid rafts [99,227] and cholesterol synthesis inhibitors reduce PrP^{Sc} formation [102,228,229], protect neurons against prion-induced synapse damage [230] and delayed the progression of experimental scrapie infections [103,104].

Many neuronal processes are affected by cholesterol depletion; cholesterol concentrations affect the sensitivity of neurotransmitter receptors [29,138,231] and synaptic vesicle recycling [136,137]. Consequently, inhibitors of cholesterol synthesis are viewed as crude pharmacological tools, so that inhibition of cholesterol synthesis is not a feasible approach to the disruption of Aβ-induced neurodegeneration. However, rafts exist as multiple heterogeneous subsets containing many different proteins and with different functions [10,232] and it is possible that compounds that specifically alter the function of those rafts involved in $A\beta_{42}$-induced neurodegeneration may be discovered.

Glimepiride

The factors that affect the formation and function of lipid rafts are not fully understood. As one of the factors that promotes the formation of lipid rafts is the presence of GPI anchors [13], the composition and function of lipid rafts might be affected by drugs that affect GPI anchors. An example of this is the binding of insulin to the insulin receptor, which activates a GPI-phospholipase C (GPI-PLC) [233-235], reduces the concentration of cholesterol in membranes [236] and alters the distribution of lipid raft-resident proteins [24]. Glimepiride, a sulphonylurea that is used to treat diabetes, mimics some of the effects of insulin upon cells, including the activation of GPI-PLC and changes in membrane biochemistry.

Glimepiride specifically affected cholesterol and lipid raft components within synapses, without affecting total cholesterol concentrations in neurons. It reduced the amounts of PrP^C in synapses, but not the expression of Thy-1, caveolin, synaptophysin or VAMP-1, an observations that is consistent with the hypothesis that lipid rafts exist as multiple, heterogeneous subsets [10] and that glimepiride affects only a specific subset. Its selectivity can be explained by its mechanism of action, it only affected lipid rafts that are precipitated by the presence of GPI-anchored proteins.

Although glimepiride reduced the expression of PrP^C [236], which acts as a receptor for $A\beta_{42}$ [201,207], it did not reduce the binding of $A\beta_{42}$ to neurons, rather glimepiride reduced the targeting of $A\beta_{42}$ to lipid rafts. In control neurons 80% of $A\beta_{42}$ was found within rafts [75-77], whereas in glimepiride-treated neurons less than 30% was in lipid rafts. In control synaptosomes $A\beta_{42}$ triggered an increase in cholesterol concentrations but had no effect on those treated with glimepiride. Lipid rafts are dynamic [10] suggesting that $A\beta_{42}$ oligomer-mediated cross-linkage of PrP^C [237] triggers an increase in cholesterol and consequently affects raft/complex formation and function. Thus, glimepiride also reduced the $A\beta_{42}$-induced activation of $cPLA_2$ in synapses. The results suggest that $A\beta_{42}$ binds to glimepiride-treated synapses by receptors that do not trigger the release of cholesterol, nor form a platform conducive to the activation of $cPLA_2$. Consequently glimepiride protects neurons against $A\beta_{42}$-induced synapse damage. As glimepiride is already used clinically and can cross the blood-brain barrier, it should be considered as a novel adjunctive treatment that may reduce the progression of AD.

Specific GPIs

GPIs have complex roles including the regulation of membrane composition, cell signalling and protein trafficking [238]. GPI anchors contain a conserved core whereby the carboxyl terminus of the protein is connected by an amide linkage to ethanolamine phosphate, mannose, glucosamine and phosphatidylinositol (PI) [239,240]. However, many variations of this core structure exist [15,240] and the composition of the GPI anchor affects the properties of the GPI-anchored proteins. GPI-anchored proteins are surrounded by lipid shells that are dependent on the sum of multiple, protein-lipid, glycan-lipid and lipid-lipid interactions [241]. Thus the composition of a GPI anchor affects the composition of the underlying membrane and consequently the function of each lipid raft. An example of this selectivity is that Thy-1 and PrP^C have compositionally different GPI anchors [242,243] and consequently occupy separate lipid rafts in neuronal membranes [244]. These rafts consist of different lipids and gangliosides [245] and PrP^C and Thy-1 traffic via different pathways [208]. Other studies showed that it is the composition of GPI anchors that directs proteins to specific lipid rafts in the absence of interactive external domains [246-248] and that some GPIs selectively inhibited the function of some GPI-anchored proteins [246]. Collectively these observations suggest that particular GPIs may be developed to target the lipid rafts responsible for prion and Aβ-mediated toxicity.

GPIs derived from PrP^C displaced PrP^C from within lipid rafts and reduced the expression of PrP^C [249]. The GPI attached to PrP^C has been reported to contain high amounts of sialic acid, galactose and mannose [242] (Figure 2). Consistent with the idea that specific GPIs inhibit the function of GPI-anchored proteins [246], PrP^C-derived GPIs reduced PrP^C expression in neurons without affecting the expression of other lipid raft proteins such as Thy-1, caveolin or VAMP-1. This effect of the PrP^C-derived GPIs was dependent upon its composition, as GPIs derived from Thy-1, CD55, monoacylated GPI or desialylated GPIs did not affect PrP^C expression.

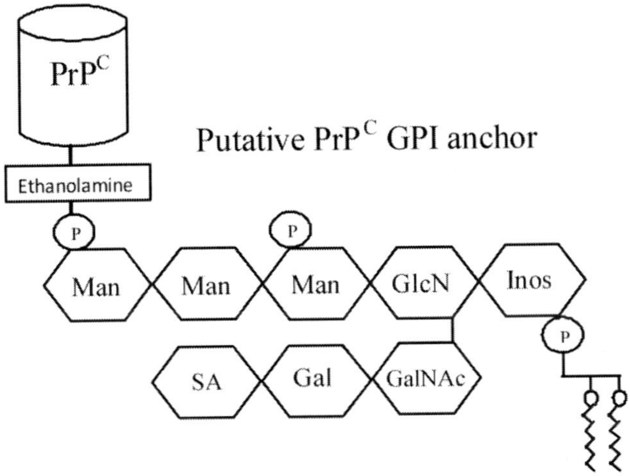

Figure 2. Schematic showing the structure of the GPI anchor attached to PrP^C. Glycan residues shown include mannose (Man), sialic acid (SA), galactose (Gal), N-acetyl galactosamine (GalNAc) and glucosamine (GlcN) as well as phosphate (P).

GPIs Reduce Prion Formation

A report that transgenic mice making anchorless PrP^C produced high amounts of infectious PrP^{Sc} without developing clinical scrapie [250] ignited interest in the role of GPI anchors in prion diseases. Many GPI-anchored proteins display distinctive recycling pathways [52,251] and PrP^C is thought to be converted into PrP^{Sc} within one of the recycling compartments [252]. Consequently any treatment that alters PrP^C expression and/or intracellular trafficking can affect the production of PrP^{Sc}. Treatment of prion-infected neuronal cell lines with GPIs reduced PrP^{Sc} production [249]. PrP^{Sc} production was reduced by specific GPIs; the GPIs isolated from PrP^C reduced PrP^{Sc} formation at significantly lower concentrations than GPIs isolated from Thy-1, CD55, or from monoacylated or desialylated GPIs. Collectively these results suggest that specific GPIs can be used to inhibit the function of lipid rafts involved in the formation of PrP^{Sc}.

GPIs in Synapse Damage

The oligomerization of GPI-anchored proteins can trigger the formation of a signalling platform indicating that one key to cell activation is the density of specific GPIs. This hypothesis is consistent with the finding that cross-linkage of PrP^C by $A\beta_{42}$ oligomers, or by mAbs increased synaptic cholesterol, activated $cPLA_2$ and triggered synapse damage [201]. The clustering of GPIs attached to PrP^C triggered activation of $cPLA_2$ and synapse damage depended upon the structure of the GPI anchor [95] indicating that the composition of GPIs modifies the signalling platform. These reports support the hypothesis that $A\beta_{42}$ oligomer-mediated cross-linkage of PrP^C triggers the release of cholesterol and the formation of a signalling platform in which activated $cPLA_2$ leads to synapse degeneration.

PrP^C-derived GPIs did not affect the binding of $A\beta_{42}$ to synapses; rather they reduced the amounts of $A\beta_{42}$ within lipid rafts. They also blocked the $A\beta_{42}$-induced rise in the concentration of cholesterol and the activation of synaptic $cPLA_2$ in synaptosomes, and protected neurons against $A\beta_{42}$-induced synapse damage. GPIs derived from PrP^C were about 100 times more efficacious than GPIs derived from Thy-1. Furthermore, the removal of sialic acid from PrP^C-derived GPIs reduced their efficacy by 100 fold, indicating that the synapse protection depended upon GPI structure. Complexes that formed between $A\beta_{42}$, PrP^C and $cPLA_2$ were sensitive to disruption by PrP^C-derived GPIs. Thus, even when added after the $A\beta_{42}$ oligomers, the PrP^C-derived GPIs dispersed $A\beta_{42}$-PrP^C complexes and significantly reduced synapse damage. Such observations suggest that specific GPIs are directed to rafts involved in $A\beta$-induced synapse damage and consequently could prove useful in the treatment of AD.

How Do We Turn Raft-Dependent Cell-Signalling off?

An inhibitory feedback pathway, whereby the production of a signalling mediator leads to suppression of its own production, is a common concept in cell regulation. Notably, PGE_2 produced after Aβ-induced activation of $cPLA_2$ increased the amounts of cyclic adenosine monophosphate (cAMP) in synapses and inhibited the activation of $cPLA_2$ (manuscript in preparation). The observation that cAMP analogues reduced the $A\beta_{42}$-induced activation of $cPLA_2$ and the production of PGE_2 in synapses demonstrated an inhibitory feedback system. Consequently, the addition of cAMP analogues protected neurons against $A\beta_{42}$-induced synapse damage.

The concept of feedback pathways also applies to the formation/ stabilisation of signalling platforms. Ligand-induced coalescence of lipid rafts to form a signalling platform results in cell activation. While the molecular processes involved in the formation of raft platforms, including the recruitment of proteins from non-raft membranes, have been widely reported, the mechanisms by which signalling platforms disperse are poorly understood. For example, the mechanisms by which $cPLA_2$ gets deactivated and returns to the cytoplasm is not well described. As the dissociation of raft platforms may control the intensity of cell stimulation, the possibility that pathology occurs when a signalling platforms fail to disperse was considered. Notably in prion diseases the self-aggregation of PrP^{Sc} generates complexes that trap signalling molecules within a highly stable platform. Similarly, the aggregation of PrP^{C} by $A\beta_{42}$ oligomers may stabilise signalling platforms [201,253]. The conditions in which cell signalling platforms fail to dissociate may lead to sustained cell activation, cell disruption and disease.

Role of the Cholesterol Cycle in Neurodegeneration

The cholesterol/cholesterol ester cycle maintains a steady level of cholesterol in cell membranes [88,89] and consequently affects the formation and dissociation of cell signalling complexes. This cycle involves the esterification of cholesterol is by ACAT and the hydrolysis of cholesterol esters by cholesterol ester hydrolases (CEH) [89] (Figure 3). ACAT inhibitors have been proposed to be treatments for AD, as they reduced the production of

$A\beta_{42}$ [90,91,254]. However, the role of ACAT in neurodegenerative diseases is complicated as it may affect different facets of pathogenesis. For example, ACAT inhibitors were 50 times more toxic to prion-infected neuronal cell lines than to their uninfected equivalents [255]. This toxicity of ACAT inhibitors was related to the increased cholesterol content of infected cells, increased activation of $cPLA_2$ and PGE_2 production [98]. Furthermore PLA_2 inhibitors blocked the cytotoxic effects of ACAT inhibitors on prion-infected cells, indicating that cholesterol esterification is an important cellular response that reduces PrP^{Sc}-induced activation of PLA_2 and protects the prion-infected ScGT1 cells.

Synapses are enriched in cholesterol [135], cholesterol esters, ACAT and CEH. The $A\beta_{42}$ oligomer-induced increase in the concentration of cholesterol in synaptosomes is due to its release from stores of esterified cholesterol. Conversely, the esterification of cholesterol limited the $A\beta_{42}$-induced increase of cholesterol in synapses suggesting that a local cholesterol deacylation/reacylation cycle controls the formation and dissociation of signalling platforms in synapses. Similarly the formation of PrP^C-$A\beta_{42}$ complexes in synapses was also regulated by the cholesterol deacylation/reacylation cycle. Inhibition of CEH reduced the formation, or enhanced the dissociation of PrP^C-$A\beta_{42}$ complexes, while inhibition of ACAT stabilised $A\beta_{42}$:PrP^C complexes.

The cholesterol cycle also affects the $A\beta_{42}$-induced activation of $cPLA_2$ in synapses; inhibition of CEH reduced, while inhibition of ACAT enhanced, $A\beta_{42}$-induced activation of $cPLA_2$, consistent with the hypothesis that esterification of cholesterol turns off $A\beta_{42}$-induced activation of $cPLA_2$. More specifically, inhibition of CEH/ACAT affected both the amount and time that $cPLA_2$ spent within lipid rafts. These pharmacological studies suggest the existence of a feedback mechanism mediated by the modification of cholesterol concentrations and the dissociation of PrP^C-$A\beta_{42}$ complexes. These observations are consistent with the hypothesis that $A\beta_{42}$ stimulates the hydrolysis of cholesterol ester stores and the release of cholesterol that stabilises lipid raft complexes in which $cPLA_2$ is activated [13].

It is thought that PrP^C-$A\beta_{42}$ complexes are transported to a cellular compartment in which ACAT esterifies cholesterol. Loss of cholesterol leads to the dissociation of PrP^C-$A\beta_{42}$ complexes and consequently the deactivation of $cPLA_2$. Thus, the inhibition of CEH reduced the $A\beta_{42}$-induced rise in cholesterol, the formation of PrP^C-$A\beta_{42}$ complexes and activation of $cPLA_2$. By contrast, inhibition of ACAT stabilised PrP^C-$A\beta_{42}$ complexes and increased activation of $cPLA_2$. As $A\beta_{42}$-induced activation of $cPLA_2$ was reduced by inhibition of CEH it was not surprising to find that inhibition of CEH also reduced $A\beta_{42}$-induced synapse damage and that inhibition of ACAT significantly increased $A\beta_{42}$-induced synapse damage.

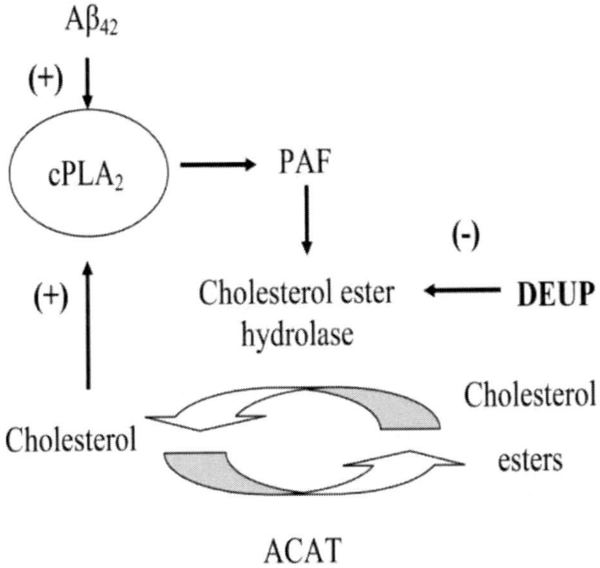

Figure 3. Feedback inhibition of $cPLA_2$ activation by cholesterol - Cartoon showing the proposed inhibitory feedback loop thought to regulate activation of $cPLA_2$ at the synapse. The presence of Aβ42 activates $cPLA_2$. The products of activated $cPLA_2$ include PAF, which activates CEH resulting in the release of cholesterol from cholesterol esters. Cholesterol is thought to stabilise lipid rafts and enhance the activation of $cPLA_2$. The esterification of cholesterol by ACAT reduces cholesterol concentrations in the membrane and leads to the dissociation of lipid raft complexes and deactivation of $cPLA_2$.

Lyso-Phospholipids Inhibit Aβ-Induced Activation of $cPLA_2$ by Dispersing Lipid Rafts

Activation of $cPLA_2$ causes the production of lyso-phospholipids (LPLs) that alter the curvature of membranes [256], and play a key role in endo and exocytosis [257-259]. LPLs are short lived within cell membranes, as they are rapidly reacylated by LPL acyltransferases (LPATs) [260,261] that restore membrane structure. The exocytosis of neurotransmitters and the retrieval of synaptic vesicles indicate that synaptic membranes are constantly changing. The synapse is rich in both PLA_2s [120,141] and LPATs [146,262,263], enzymes that regulate the ratio of phospholipid to LPL and hence control membrane shape and function. Notably, endophilin 1, an LPAT [146] that is concentrated within the synapse, is required for the formation of synaptic vesicles [146,262,263].

As synthetic LPLs target lipid rafts [264-267], their effects on $A\beta_{42}$-induced cell signalling and synapse damage were explored, using CI-976, an inhibitor of LPAT [268] that causes LPLs to accumulate in membranes [269]. CI-976 increased the $A\beta_{42}$-induced the production of LPLs and reduced the $A\beta_{42}$-induced activation of $cPLA_2$ in synapses. Similarly, the addition of LPLs reduced the $A\beta_{42}$-induced activation of $cPLA_2$. They also regulated the stability of PrP^C-$A\beta_{42}$ complexes; CI-976 reduced the formation and/or stability of PrP^C-$A\beta_{42}$ complexes and the addition of LPLs caused the rapid dissociation of PrP^C-$A\beta_{42}$ complexes. CI-976 or LPLs increased the internalisation of PrP^C-$A\beta_{42}$ complexes, observations that are consistent with the hypothesis that LPLs facilitate the endocytosis of lipid rafts [267]. The $A\beta_{42}$-induced increase in cholesterol was reduced in synaptosomes pre-treated with CI-976, or treated with LPLs, results consistent with reports that LPLs affect cholesterol transport [270], cholesterol levels [271] and alter the organisation of rafts [266]. Consequently, pre-treatment with C1-976, or the addition of LPLs, protected neurons against $A\beta_{42}$-induced synapse damage.

Nutraceuticals that May Preserve Synapse Health

Epidemiological studies support the idea that dietary factors can affect the progression of neurodegenerative diseases such as Alzheimer's and Parkinson's diseases. More specifically, there is burgeoning interest, both academic and commercial, in the role of nutraceuticals; foods that provide medical or health benefits. The assay systems developed in the laboratory provide simple ways of testing their effects on synapses.

Ethanol

Some epidemiological studies suggest that moderate consumption of alcohol protects against the development of AD and PD [272-274]. Since ethanol readily crosses the BBB, physiologically relevant concentrations (0.02

- 0.08%) were tested on neurons. Exposure of neurons to 0.04% ethanol (the equivalent to a large glass of wine) protected neurons against Aβ and αSN-induced synapse damage [275]. It did not affect the accumulation of Aβ at synapses, rather it reduced Aβ and αSN-induced activation of $cPLA_2$ within synapses. Reports that ethanol affects cholesterol distribution in neurons [276] raised the possibility that it alters the lipid rafts in which $A\beta_{42}$ activates $cPLA_2$.

Ethanol did not affect the targeting of $A\beta_{42}$ to lipid rafts; rather it altered the distribution of $cPLA_2$ in synapses. Upon cell stimulation, $cPLA_2$ migrates to specific intracellular membranes [195] and the addition of $A\beta_{42}$ to control synapses triggered the migration of $cPLA_2$ from the cytoplasm to lipid rafts [10]. In contrast, less $cPLA_2$ was found in lipid rafts in synaptosomes that had been pre-treated with ethanol and incubated with $A\beta_{42}$. That ethanol protects neurons against synapse damage induced by $A\beta_{42}$ and αSN further supports the hypothesis that moderate alcohol consumption protects against the dementia that develops in patients with AD or PD.

Caffeine

Caffeine is a broad spectrum inhibitor of phosphodiesterases (PDE)s, the enzymes that metabolise cAMP and are responsible for terminating its action. As cAMP inhibits the activation of $cPLA_2$ in synapses the presence of caffeine enhanced the $A\beta_{42}$-induced increase in cAMP. This resulted in reduced activation of $cPLA_2$ and protection against $A\beta_{42}$-induced synapse damage (Bate et al., manuscript in preparation). This observation is intriguing as a number of epidemiological studies suggest that regular coffee consumption protects against Alzheimer's and Parkinson's diseases and other dementias [277-281].

Vitamin E

There is increasing interest in the effects of vitamins on the progression of neurodegenerative diseases [282,283]. For example, vitamin E has been reported to alter Aβ levels and protect neurons [284,285], while some B vitamins are also associated with protection against neurodegeneration [286,287]. However, the effects of vitamins upon AD pathogenesis are controversial [288]. Vitamin E (α-tocopherol) protected neurons against $A\beta_{42}$-induced synapse damage in neuronal cultures (manuscript in preparation]. It reduced both $A\beta_{42}$ and prion-induced activation of $cPLA_2$ in synapses, consistent with reports that vitamin E inhibits PLA_2 in other tissues [289-291] and suppresses PGE_2 production [292]. It did not directly inhibit $cPLA_2$ rather it reduced the Aβ-induced translocation of $cPLA_2$ to lipid rafts. Since vitamin E lowers cholesterol through an anti-oxidant independent mechanism [293] we conclude that it altered rafts by preventing the stabilisation and activation of $cPLA_2$. This hypothesis is consistent with the emerging concept that the sequestration of signalling enzymes away from their substrates helps regulate signalling pathways.

Sphingolipids

The sphingolipids constitute one of the major components of lipid rafts; they have significant effects on the formation and function of lipid rafts as they contain saturated fatty acids that help solubilise cholesterol. In particular, gangliosides have been implicated in the structure of Aβ [294], the progression of AD [295] and prion formation [296]. The metabolism of sphingolipids generates biologically active lipid mediators that regulate the formation and/or stability of rafts [297-302]. Notably activation of the $cPLA_2$ cascade and sphingolipid metabolism are interrelated; PAF modulates sphingolipid metabolism [303] and sphingolipids regulate the activation of $cPLA_2$ [304,305]. Dysregulation of sphingomyelinase is observed in AD [306] and the cellular prion protein is involved in sphingolipid-mediated cell signalling [307]. However, this subject is sufficiently complex to warrant a chapter in its own right and consequently is not discussed here.

Oxidation

GPI-anchored proteins and sphingolipids contain predominantly saturated acyl chains that increase cholesterol solubility and lipid raft formation. Oxidation of acyl chains results in unsaturated fatty acids that reduce cholesterol solubility; consequently oxidation of cell membranes may affect the formation and function of lipid rafts. Oxidative stress-induced abnormalities in cholesterol and ceramide metabolism have been reported in AD [308].

Polyunsaturated Fatty Acids (PUFA)

PUFA have such extensive effects upon cell signalling that only a brief overview can be mentioned here. Feeding cells PUFA leads to unsaturated fatty acids replacing saturated ones and can have two distinct effects on lipid rafts and cell signalling. Firstly, the metabolism of PUFA can generate potent bioactive lipids, and secondly the presence of PUFA can alter the biochemistry of cell membranes and regulate lipid raft formation. PUFAs such as arachidonic and docosahexaenoic acids can be metabolised to a variety of metabolites, each binding to specific receptors and having different actions. PUFAs also have profound effects upon cell membranes; they affect the solubility of cholesterol, the formation of lipid rafts and cell signalling within lipid rafts. For example, the increased consumption of polyunsaturated fatty acids such as docosahexaenoic acid is thought to protect against dementia [309,310]. Neurons cultured with docosahexaenoic acid were resistant to Aβ and prion-mediated synapse damage [311]. Protection was associated with altered trafficking of Aβ in neurons and reduced accumulation of Aβ in synapses.

Monoacylated PrP^C

The hypothesis that the cross-linkage of PrP^C, whether by $A\beta_{42}$ oligomers or by specific mAbs, triggers the formation of a raft complex that leads to the recruitment and activation of $cPLA_2$ depends upon the targeting of PrP^C to lipid rafts. The nature of lipid anchors on a protein determines its localisation within the plasma membrane [312] and the composition of GPI anchors direct proteins to specific rafts [246-248]. Since PrP^C-mediated cell signalling and synapse damage depends on the composition of the GPI anchor [95], it seemed likely that the modification of its GPI anchor could affect the function of PrP^C. Monoacylated PrP^C was prepared from PrP^C by digestion with PLA_2 which allowed the examination of modifications of the GPI anchor that could not be achieved by genetic manipulation methods (Figure 4). Whereas PrP^C is a recycling protein [313,314] with a half-life of less than 24 hours, monoacylated PrP^C was more stable at the cell surface and had a half-life of more than 5 days. The nature and number of the acyl chains contained within GPI anchors are factors that affect lipid raft formation and consequently monoacylated PrP^C was found within the normal cell membrane rather than within lipid rafts [315].

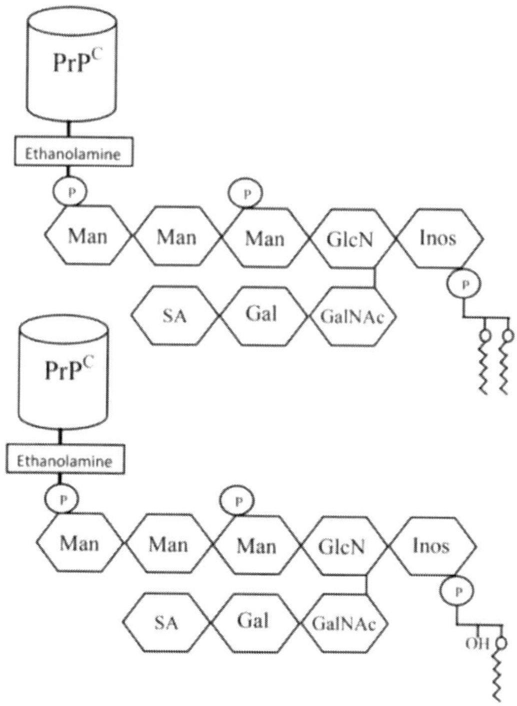

Figure 4. Schematic comparing PrP^C with monoacylated PrP^C. Glycan residues shown include mannose (Man), sialic acid (SA), galactose (Gal), N-acetyl galactosamine (GalNAc) and glucosamine (GlcN) as well as phosphate (P).

Monoacylated PrP^C in Prion Diseases

Although PrP^C is essential for PrP^{Sc} formation [316,317], not all cells that express PrP^C permit PrP^{Sc} replication. As efficient PrP^{Sc} formation requires the targeting of PrP^C to specific membranes [318] the factors that affect the intracellular trafficking of PrP^C are critical in determining PrP^{Sc} replication. Not only was monoacylated PrP^C not converted into PrP^{Sc} it also reduced PrP^{Sc} formation [315]. The integrity of lipid rafts is thought to be necessary for PrP^{Sc} formation. Both the composition and function of lipid rafts are controlled by a dynamic "induced fit" model [10] and the binding of monoacylated PrP^C to PrP^{Sc} modifies those lipid rafts that are involved in PrP^{Sc} formation. Since monoacylated PrP^C does not sequester cholesterol, the membrane surrounding a complex between PrP^{Sc} and monoacylated PrP^C contained less cholesterol than membranes surrounding PrP^{Sc} and PrP^C. As the cholesterol content of cell membranes is critical for the formation of lipid rafts [13] the binding of monoacylated PrP^C to PrP^{Sc} reduced the cholesterol content of local membranes to a level below that required for the conversion of PrP^C to PrP^{Sc}.

Lipid rafts act as domains in which GPI anchors attached to PrP^C interact with cell signalling pathways [30] and it is the clustering of GPI anchors attached to PrP proteins that leads to the activation of $cPLA_2$. The presence of monoacylated PrP^C reduced the activation of $cPLA_2$ within prion-infected cells as it caused the dissociation of $cPLA_2$ from PrP^{Sc}-containing lipid rafts. The targeting of $cPLA_2$ to membranes containing their substrates can regulate cell signalling, including the formation of PAF that facilitate PrP^{Sc} formation [168]. Collectively, these observations suggest that while PrP^C binds to PrP^{Sc} where it activates the $cPLA_2$ that facilitates the conversion of PrP^C to PrP^{Sc}, the binding of monoacylated PrP^C to PrP^{Sc} changed the composition of the underlying membrane so that it inhibited $cPLA_2$ and prevented PrP^{Sc} formation.

Monoacylated PrP^C in AD

Neurons loaded with monoacylated PrP^C bound more $A\beta_{42}$ than control neurons and complexes formed between monoacylated PrP^C and $A\beta_{42}$ were more stable than those formed between PrP^C and $A\beta_{42}$. It was therefore surprising that neurons decorated with monoacylated PrP^C were protected against $A\beta_{42}$-induced synapse damage. Whereas the cross-linkage of PrP^C by $A\beta_{42}$ oligomers leads to the clustering of specific GPI anchors, activation of $cPLA_2$ and synapse damage [201], the presence of monoacylated PrP^C reduced $A\beta_{42}$-induced activation of $cPLA_2$ in synapses.

Most $A\beta_{42}$ was found outside lipid rafts in neurons decorated with monoacylated PrP^C. Since activation of $cPLA_2$ occurs within lipid rafts [199,201], the presence of monoacylated PrP^C resulted in the sequestration of $A\beta_{42}$ into non-signalling membrane domains. The targeting of proteins to lipid rafts also affects the subsequent trafficking of proteins [319] and consequently when monoacylated PrP^C sequestered $A\beta_{42}$ in non-raft membranes at the cell body it prevented the trafficking of $A\beta_{42}$ to synapses. Thus, monoacylated PrP^C acts as a molecular sponge and adsorbs $A\beta_{42}$ to specific cell compartments, so preventing $A\beta_{42}$ binding to PrP^C and generating the raft dependent cell signalling that leads to synapse damage. The toxic effects of $A\beta$ oligomers are mimicked by the cross-linkage of PrP^C with mAbs [201]. Cross-linkage of monoacylated PrP^C did not affect the cholesterol content of cell membranes; indicating that two saturated fatty acids must be contained within the GPI anchor to sequester cholesterol and precipitate the formation of lipid rafts [21]. In addition, the cross-linkage of monoacylated PrP^C by mAbs did not activate $cPLA_2$ nor did it trigger synapse damage.

Role of Lipid Rafts in the Clearance of Neurotoxins

The chronic nature of neurodegenerative diseases is consistent with the hypothesis that it is the accumulation of neurotoxins that triggers neurodegeneration. In AD neurodegeneration is preceded by intraneuronal accumulation of $A\beta_{42}$ [320,321] whereas in prion diseases it is PrP^{Sc} or the disease-associated PrP82-146 peptide that accumulate [322]. The accumulation of $A\beta_{42}$ or PrP82-146 within the brain may result from their slow rate of degradation within neurons. Both PrP82-146 and $A\beta_{42}$ can be degraded, $A\beta_{42}$ is digested by proteases including neprilysin [323], insulysin [324], cathepsin B [325] and acyl peptide hydrolase [326] and both $A\beta_{42}$ and PrP82-146 were rapidly degraded in microglial cells, suggesting it should be possible to increase their degradation within neurons.

Observations that $A\beta_{42}$ oligomers and PrP peptides are targeted to lipid rafts [76-78,227] may be of significance since many raft-associated proteins traffic through cells by way of recycling pathways that avoid the lysosomes [327-329]. Both $A\beta_{42}$ oligomers and PrP peptides are slowly degraded in cultured neurons [330] suggesting that the targeting of $A\beta_{42}$/PrP82-146 to lipid rafts may contribute to their accumulation within neurons. The trafficking and hence the metabolism of $A\beta_{42}$/PrP82-146 is sensitive to pharmacological manipulation; treatment of neurons with the cholesterol synthesis inhibitor squalestatin [331] affected the distribution of $A\beta_{42}$/PrP82-146. It reduced the amounts of $A\beta_{42}$/PrP82-146 found within lipid rafts and increased amounts of $A\beta_{42}$/PrP82-146 in the lysosomes; consequently the half-lives of PrP82-146 and $A\beta_{42}$ were reduced from 5 days to less than 24 hours. Similar results were obtained with neurons treated with simvastatin, a clinically approved HMG-CoA reductase inhibitor.

Reports that Aβ affects cholesterol homeostasis [80-82] suggested that it modifies the biochemistry of cell membranes, creating the rafts that are responsible for its trafficking. $A\beta_{42}$ activates $cPLA_2$ and the production of PAF [151,155] and. PAF receptor antagonists also reduced the amounts of PrP82-146 and $A\beta_{42}$ in lipid rafts [330]. PAF is involved in endocytosis and endosome formation [332,333]. PAF antagonists increased the amounts of

PrP82-146 and $A\beta_{42}$ entering the lysosomes, sites of protein degradation [334], resulting in half-lives of less than 24 hours. These results support the hypothesis that PrP and $A\beta_{42}$-induced activation of PLA_2 leads to the production of PAF, which regulates raft formation. In normal cells PrP and $A\beta_{42}$ are stabilised in lipid rafts which traffic via a recycling pathway that avoids the lysosomes and hence peptide digestion. PAF antagonists destabilise these rafts, redirecting PrP and $A\beta_{42}$ into the lysosomes so increasing their digestion. The novel observation that the activation of specific signalling pathways controls the trafficking and degradation of $A\beta_{42}$ may contribute towards the development of drugs that will delay the progression of AD.

Conclusion

I have attempted to show how the accumulation of specific, aggregated peptides causes pathological cell signalling that leads to the dysfunction and degeneration of synapses. The consequences of synapse degeneration are the loss of memory and the progressive dementia that characterises Alzheimer's and prion diseases. The pathological signalling depends upon lipid raft-associated proteins or proteins that are recruited into lipid rafts. I propose that critical concentrations of these peptides trigger the formation of a signalling platform and it is aberrant cell signalling within these platforms that leads to synapse damage.

It is hoped that an understanding of the processes involved in the formation, composition and dissociation of such signalling platforms can lead to the discovery of novel therapeutics that protect synapses. They would not be based upon conventional enzyme inhibitors but on factors that affect the membrane micro-environment within which signalling platforms are generated. Studies on cholesterol synthesis inhibitors led the way to understanding these processes, concluding that reducing cholesterol levels within cell membranes disrupted cell signalling. However as cholesterol effects many cellular processes compounds that target pathological lipid raft signalling more specifically have been sought. Here I describe a number of compounds that may be useful in disrupting disease-associated lipid rafts in synapses. While many of the compounds mentioned in this chapter have pharmacokinetic limitations, I hope to establish the principle that some may be capable of disrupting specific lipid rafts involved in disease progression.

References

[1] Lingwood, D. & Simons, K. (2007). *Nat. Protocols*, *2*, 2159-2165

[2] Williamson, R., Thompson, A., Abu, M., Hye, A., Usardi, A., Lynham, S., Anderton, B. & Hanger, D. (2010). *BMC Neuroscience*, *11*, 120

[3] Chen, X., Jen, A., Warley, A., Lawrence, M. J., Quinn, P. J. & Morris, R. J. (2009). *Biochem J*, *417*, 525-533

[4] Macdonald, J. L. & Pike, L. J. (2005). *Journal of Lipid Research*, *46*, 1061-1067

[5] Smart, E. J., Ying, Y. S., Mineo, C. & Anderson, R. G. (1995). *Proc Natl Acad Sci U S A*, *92*, 10104-10108

[6] Morris, R. J., Jen, A. & Warley, A. (2011). *Journal of Neurochemistry*, *116*, 671-677

[7] Simons, K. & Gerl, M. J. (2010). *Nat Rev Mol Cell Biol*, *11*, 688-699

[8] Mukherjee, S., Zha, X., Tabas, I. & Maxfield, F. R. (1998). *Biophys J*, *75*, 1915-1925

[9] Pike, L. J. (2006). *J Lipid Res*, *47*, 1597-1598

[10] Pike, L. J. (2004). *Biochem J*, *378*, 281-292

[11] Ohmi, Y., Ohkawa, Y., Yamauchi, Y., Tajima, O. & Furukawa, K. (2012). *Neurochem Res*, *37*, 1185-1191

[12] Cantu, L., Del Favero, E., Sonnino, S. & Prinetti, A. (2011). *Chem Phys Lipids*, *164*, 796-810

[13] Brown, D. A. & London, E. (2000). *J.Biol.Chem.*, *75*, 17221-17224

[14] Varma, R. & Mayor, S. (1998). *Nature*, *394*, 798-801

[15] Major, S. & Reizman, H. (2004). *Nature Reviews (Mol Cell Biol).*, *5*, 110 -120

[16] Legler, D. F., Doucey, M. A., Schneider, P., Chapatte, L., Bender, F. C. & Bron, C. (2005). *FASEB J.*, *19*, 73-75

[17] Mayor, S. & Riezman, H. (2004). *Nat Rev Mol Cell Biol*, *5*, 110-120

[18] Levental, I., Lingwood, D., Grzybek, M., Coskun, Ü. & Simons, K. (2010). *Proc Natl Acad Sci U S A*, *107*, 22050-22054

[19] Ikonen, E. (2001). *Curr Opin Cell Biol*, *13*, 470-477

[20] Schroeder, F., Gallegos, A. M., Atshaves, B. P., Storey, S. M., McIntosh, A. L., Petrescu, A. D., Huang, H., Starodub, O., Chao, H., Yang, H., Frolov, A. & Kier, A. B. (2001). *Exp. Biol. Med*, *226*, 873-890

[21] Schroeder, R., London, E. & Brown, D. (1994). *Proc.Natl.Acad.Sci. U.S.A*, *91*, 12130-12134

[22] Maeda, Y., Tashima, Y., Houjou, T., Fujita, M., Yoko-o, T., Jigami, Y., Taguchi, R. & Kinoshita, T. (2007). *Mol.Biol.Cell*, *18*, 1497-1506

[23] Huang, J. & Feigenson, G. W. (1999). *Biophys J*, *76*, 2142-2157

[24] Müller, G., Schulz, A., Wied, S. & Frick, W. (2005). *Biochemical Pharmacology*, *69*, 761-780

[25] Lingwood, D., Kaiser, H. J., Levental, I. & Simons, K. (2009). *Biochem Soc Trans*, *37*, 955-960

[26] Rothberg, K. G., Ying, Y. S., Kamen, B. A. & Anderson, R. G. (1990). *J Cell Biol*, *111*, 2931-2938

[27] Bilderback, T. R., Grigsby, R. J. & Dobrowsky, R. T. (1997). *Journal of Biological Chemistry*, *272*, 10922-10927

[28] Baenziger, J. E., Morris, M. L., Darsaut, T. E. & Ryan, S. E. (2000). *Journal of Biological Chemistry*, *275*, 777-784

[29] Sooksawate, T. & Simmonds, M. A. (2001). *Neuropharmacology*, *40*, 178-184

[30] Simons, K. & Toomre, D. (2000). *Nat Rev Mol Cell Biol*, *1*, 31-39

[31] Li, S., Couet, J. & Lisanti, M. P. (1996). *J. Biol. Chem*, *271*, 29182-29190

[32] Gorodinsky, A. & Harris, D. A. (1995). *J.Cell Biol.*, *129*, 619-627

[33] Cooper, D. M. F. & Crossthwaite, A. J. (2006). *Trends Pharm Sci*, *27*, 426-431

[34] Oh, P. & Schnitzer, J. E. (2001). *Mol.Biol.Cell*, *12*, 685-698

[35] Pike, L. J. & Miller, J. M. (1998). *J.Biol.Chem.*, *273*, 22298-22304

[36] Prior, I. A., Harding, A., Yan, J., Sluimer, J., Parton, R. G. & Hancock, J. F. (2001). *Nat Cell Biol*, *3*, 368-375

[37] Graziani, A., Bricko, V., Carmignani, M., Graier, W. F. & Groschner, K. (2004). *Cardiovasc.Res*, *64*, 234-242

[38] Bate, C., Rumbold, L. & Williams, A. (2007). *Journal of Neuroinflammation*, *4*, 5

[39] Jin, S., Zhou, F., Katirai, F. & Li, P. L. (2011). *Antioxidants & redox signaling*, *15*, 1043-1083

[40] Mayor, S. & Rao, M. (2004). *Traffic*, *5*, 231-240

[41] Lingwood, D., Ries, J., Schwille, P. & Simons, K. (2008). *Proc Natl Acad Sci USA*, *105*, 10005-10010

[42] Lingwood, D. & Simons, K. (2010). *Science*, *327*, 46-50

[43] Hancock, J. F. (2006). *Nat Rev Mol Cell Biol*, *7*, 456-462

[44] Helms, J. B. & Zurzolo, C. (2004). *Traffic*, *5*, 247-254

[45] Harder, T., Scheiffele, P., Verkade, P. & Simons, K. (1998). *J Cell Biol*, *141*, 929-942

[46] Engelman, D. M. (2005). *Nature*, *438*, 578-580

[47] Hammond, A. T., Heberle, F. A., Baumgart, T., Holowka, D., Baird, B. & Feigenson, G. W. (2005). *Proc Natl Acad Sci U S A*, *102*, 6320-6325
[48] Ewers, H., Romer, W., Smith, A. E., Bacia, K., Dmitrieff, S., Chai, W., Mancini, R., Kartenbeck, J., Chambon, V., Berland, L., Oppenheim, A., Schwarzmann, G., Feizi, T., Schwille, P., Sens, P., Helenius, A. & Johannes, L. (2010). *Nat Cell Biol*, *12*, 11-18
[49] Gri, G., Molon, B., Manes, S., Pozzan, T. & Viola, A. (2004). *Immunol Lett*, *94*, 247-252
[50] Eisenberg, S., Shvartsman, D. E., Ehrlich, M. & Henis, Y. I. (2006). *Mol. Cell. Biol.*, *26*, 7190-7200
[51] Field, K. A., Holowka, D. & Baird, B. (1995). *Proc Natl Acad Sci U S A*, *92*, 9201-9205
[52] Rajendran, L. & Simons, K. (2005). *J Cell Sci*, *118*, 1099-1102
[53] Selkoe, D. J. (2002). *Science*, *298*, 789-791
[54] Tanzi, R. E. (2005). *Nat Neurosci*, *8*, 977-979
[55] Esler, W. P. & Wolfe, M. S. (2001). *Science*, *293*, 1449-1454
[56] Selkoe, D. J. (1994). *J Neuropath Exp Neurol*, *53*, 438-447
[57] De Felice, F. G., Wu, D., Lambert, M. P., Fernandez, S. J., Velasco, P. T., Lacor, P. N., Bigio, E. H., Jerecic, J., Acton, P. J., Shughrue, P. J., Chen-Dodson, E., Kinney, G. G. & Klein, W. L. (2008). *Neurobiol Aging*, *29*, 1334-1347
[58] Lacor, P. N., Buniel, M. C., Furlow, P. W., Sanz Clemente, A., Velasco, P. T., Wood, M., Viola, K. L. & Klein, W. L. (2007). *J Neurosci*, *27*, 796-807
[59] Shankar, G. M., Bloodgood, B. L., Townsend, M., Walsh, D. M., Selkoe, D. J. & Sabatini, B. L. (2007). *J Neurosci*, 27, 2866-2875
[60] Benilova, I., Karran, E. & De Strooper, B. (2012). *Nat Neurosci*, *15*, 349-357
[61] Yankner, B. A. & Lu, T. (2009). *J.Biol.Chem.*, *284*, 4755-4759
[62] Lambert, M. P., Barlow, A. K., Chromy, B. A., Edwards, C., Freed, R., Liosatos, M., Morgan, T. E., Rozovsky, I., Trommer, B., Viola, K. L., Wals, P., Zhang, C., Finch, C. E., Krafft, G. A. & Klein, W. L. (1998). *Proc Natl Acad Sci USA*, *95*, 6448-6453
[63] Klein, W. L., Krafft, G. A. & Finch, C. E. (2001). *Trends Neurosci*, *24*, 219-224
[64] Haass, C. & Selkoe, D. J. (2007). *Nat Rev Mol Cell Biol*, *8*, 101-112
[65] Chromy, B. A., Nowak, R. J., Lambert, M. P., Viola, K. L., Chang, L., Velasco, P. T., Jones, B. W., Fernandez, S. J., Lacor, P. N., Horowitz, P.,

Finch, C. E., Krafft, G. A. & Klein, W. L. (2003). *Biochemistry*, *42*, 12749-12760
[66] Puglielli, L., Tanzi, R. E. & Kovacs, D. M. (2003). *Nat Neurosci*, *6*, 345-351
[67] Simons, M., Keller, P., Dichgans, J. & Schulz, J. B. (2001). *Neurology*, *57*, 1089-1093
[68] Bu, G. (2009). *Nat Rev Neurosci*, *10*, 333-344
[69] Jick, H., Zornberg, G. L., Jick, S. S., Seshadri, S. & Drachman, D. A. (2000). *Lancet*, *356*, 1627-1631
[70] Li, G., Larson, E. B., Sonnen, J. A., Shofer, J. B., Petrie, E. C., Schantz, A., Peskind, E. R., Raskind, M. A., Breitner, J. C. S. & Montine, T. J. (2007). *Neurology*, *69*, 878-885
[71] Neuhaus, O., Stuve, O., Zamvil, S. S. & Hartung, H. P. (2005). *CNS Drugs*, *19*, 833-841
[72] Shobab, L. A., Hsiung, G.-Y. R. & Feldman, H. H. (2005). *Lancet Neurol*, *4*, 841-852
[73] Hicks, D. A., Nalivaeva, N. N. & Turner, A. J. (2012). *Frontiers in physiology*, *3*, 189
[74] Vetrivel, K. S. & Thinakaran, G. (2010). *Biochim Biophys Acta*, *1801*, 860-867
[75] Kawarabayashi, T., Shoji, M., Younkin, L. H., Wen-Lang, L., Dickson, D. W., Murakami, T., Matsubara, E., Abe, K., Ashe, K. H. & Younkin, S. G. (2004). *J Neurosci*, *24*, 3801-3809
[76] Lee, S. J., Liyanage, U., Bickel, P. E., Xia, W., Lansbury, P. T., Jr. & Kosik, K. S. (1998). *Nat Med*, *4*, 730-734
[77] Oshima, N., Morishima-Kawashima, M., Yamaguchi, H., Yoshimura, M., Sugihara, S., Khan, K., Games, D., Schenk, D. & Ihara, Y. (2001). *Am J Pathol*, *158*, 2209-2218
[78] Williamson, R., Usardi, A., Hanger, D. P. & Anderton, B. H. (2008). *FASEB J*, *22*, 1552-1559
[79] Grimm, M. O., Grimm, H. S., Patzold, A. J., Zinser, E. G., Halonen, R., Duering, M., Tschape, J. A., De Strooper, B., Muller, U., Shen, J. & Hartmann, T. (2005). *Nat Cell Biol*, *7*, 1118-1123
[80] Grimm, M. O. W., Grimm, H. S. & Hartmann, T. (2007). *Trends Mol Med*, *13*, 337-344
[81] Liu, Y., Peterson, D. A. & Schubert, D. (1998). *Proc Natl Acad Sci U S A*, *95*, 13266-13271
[82] Yao, Z. X. & Papadopoulos, V. (2002). *FASEB J*, *16*, 1677-1679

[83] Refolo, L. M., Malester, B., LaFrancois, J., Bryant-Thomas, T., Wang, R., Tint, G. S., Sambamurti, K., Duff, K. & Pappolla, M. A. (2000). *7*, 321-331

[84] Wolozin, B., Kellman, W., Ruosseau, P., Celesia, G. G. & Siegel, G. (2000). *Arch.Neurol.*, *57*, 1439-1443

[85] Bate, C. & Williams, A. (2007). *Neuropharmacology*, *53*, 222-231

[86] Wang, S. S., Rymer, D. L. & Good, T. A. (2001). *J.Biol.Chem.*, *276*, 42027-42034

[87] Schneider, A., Schulz-Schaeffer, W., Hartmann, T., Schulz, J. B. & Simons, M. (2006). *Neurobiol Dis*, *23*, 573-577

[88] Simons, K. & Ikonen, E. (2000). *Science*, *290*, 1721-1726

[89] Chang, T. Y., Chang, C. C. Y. & Cheng, D. (1997). *Ann Rev Biochem*, *66*, 613-638

[90] Hutter-Paier, B., Huttunen, H. J., Puglielli, L., Eckman, C. B., Kim, D. Y., Hofmeister, A., Moir, R. D., Domnitz, S. B., Frosch, M. P., Windisch, M. & Kovacs, D. M. (2004). *Neuron*, *44*, 227-238

[91] Puglielli, L., Konopka, G., Pack-Chung, E., Ingano, L. A., Berezovska, O., Hyman, B. T., Chang, T. Y., Tanzi, R. E. & Kovacs, D. M. (2001). *Nat Cell Biol*, *3*, 905-912

[92] Refolo, L. M., Pappolla, M. A., LaFrancois, J., Malester, B., Schmidt, S. D., Thomas-Bryant, T., Tint, G. S., Wang, R., Mercken, M., Petanceska, S. S. & Duff, K. E. (2001). *Neurobiology of Disease*, *8*, 890-899

[93] Prusiner, S. B. (1998). *Proc Natl Acad Sci U S A*, *95*, 13363-13383

[94] Pan, K. M., Baldwin, M., Nguyen, J., Gasset, M., Serban, A., Groth, D., Mehlhorn, I., Huang, Z., Fletterick, R. J. & Cohen, F. E. (1993). *Proc Natl Acad Sci U S A*, *90*, 10962-10966

[95] Bate, C. & Williams, A. (2012). *J.Biol.Chem.*, *287*, 7935-7944

[96] Fioriti, L., Angeretti, N., Colombo, L., De Luigi, A., Colombo, A., Manzoni, C., Morbin, M., Tagliavini, F., Salmona, M., Chiesa, R. & Forloni, G. (2007). *J. Neurosci.*, *27*, 1576-1583

[97] Russelakis-Carneiro, M., Hetz, C., Maundrell, K. & Soto, C. (2004). *Am.J.Pathol.*, *165*, 1839-1848

[98] Bate, C., Tayebi, M. & Williams, A. (2008). *BMC Biol*, *6*, 8

[99] Naslavsky, N., Stein, R., Yanai, A., Friedlander, G. & Taraboulos, A. (1997). *J.Biol.Chem*, *272*, 6324-6331

[100] Sotgia, F., Razani, B., Bonuccelli, G., Schubert, W., Battista, M., Lee, H., Capozza, F., Schubert, A. L., Minetti, C., Buckley, J. T. & Lisanti, M. P. (2002). *Mol.Cell Biol.*, *22*, 3905-3926

[101] Vey, M., Pilkuhn, S., Wille, H., Nixon, R., DeArmond, S. J., Smart, E. J., Anderson, R. G., Taraboulos, A. & Prusiner, S. B. (1996). *Proc.Natl.Acad.Sci.U.S.A.*, *93*, 14945-14949
[102] Taraboulos, A., Scott, M., Semenov, A., Avrahami, D., Laszlo, L., Prusiner, S. B. & Avraham, D. (1995). *J Cell Biol*, *129*, 121-132
[103] Fai Mok, S. W., Thelen, K. M., Riemer, C., Bamme, T., Gultner, S., Lutjohann, D. & Baier, M. (2006). *Biochem Biophys Res Comm*, *348*, 697-702
[104] Kempster, S., Bate, C. & Williams, A. (2007). *Neuroreport*, *18*, 479-482
[105] Schengrund, C. -L. (2010). *Brain Research Bulletin*, *82*, 7-17
[106] Bar-On, P., Crews, L., Koob, A. O., Mizuno, H., Adame, A., Spencer, B. & Masliah, E. (2008). *J Neurochem*, *105*, 1656-1667
[107] Koob, A. O., Ubhi, K., Paulsson, J. F., Kelly, J., Rockenstein, E., Mante, M., Adame, A. & Masliah, E. (2010). *Experimental Neurology*, *221*, 267-274
[108] Ghosh, A., Roy, A., Matras, J., Brahmachari, S., Gendelman, H. E. & Pahan, K. (2009). *J. Neurosci.*, *29*, 13543-13556
[109] Becker, C. & Meier, C. R. (2009). *Expert Opinion on Drug Safety*, *8*, 261-271
[110] Counts, S. E., Nadeem, M., Lad, S. P., Wuu, J. & Mufson, E. J. (2006). *J Neuropath Exp Neurol*, *65*, 592-601
[111] Reddy, P. H., Mani, G., Park, B. S., Jacques, J., Murdoch, G., Whetsell, W., Jr., Kaye, J. & Manczak, M. (2005). *J Alzheimers Dis*, *7*, 103-117
[112] Masliah, E., Terry, R. D., Alford, M., DeTeresa, R. & Hansen, L. A. (1991). *Am J Path*, *138*, 235-246
[113] Scheff, S. W., Price, D. A., Schmitt, F. A., DeKosky, S. T. & Mufson, E. J. (2007). *Neurology*, *68*, 1501-1508
[114] Terry, R. D., Masliah, E., Salmon, D. P., Butters, N., DeTeresa, R., Hill, R., Hansen, L. A. & Katzman, R. (1991). *Ann Neurol*, *30*, 572-580
[115] DeKosky, S. T. & Scheff, S. W. (1990). *Ann Neurol*, *27*, 457-464
[116] Hamos, J. E., DeGennaro, L. J. & Drachman, D. A. (1989). *Neurology*, *39*, 355-361
[117] Sze, C. I., Troncoso, J. C., Kawas, C., Mouton, P., Price, D. L. & Martin, L. J. (1997). *J Neuropath Exp Neurol*, *56*, 933-944
[118] Heinonen, O., Soininen, H., Sorvari, H., Kosunen, O., Paljarvi, L., Koivisto, E. & Riekkinen, Sr. (1995). *Neuroscience*, *64*, 375-384
[119] Lipton, A. M., Cullum, C. M., Satumtira, S., Sontag, E., Hynan, L. S., White, C. L., 3rd & Bigio, E. H. (2001). *Arch Neurol*, *58*, 1233-1239
[120] Bate, C., Tayebi, M. & Williams, A. (2010). *Mol Neurodegener*, *5*, 13

[121] Mehta, P. D., Pirttila, T., Mehta, S. P., Sersen, E. A., Aisen, P. S. & Wisniewski, H. M. (2000). *Arch Neurol*, *57*, 100-105
[122] Bibl, M., Mollenhauer, B., Esselmann, H., Lewczuk, P., Klafki, H.-W., Sparbier, K., Smirnov, A., Cepek, L., Trenkwalder, C., Ruther, E., Kornhuber, J., Otto, M. & Wiltfang, J. (2006). *Brain* 129, 1177-1187
[123] Lue, L. F., Kuo, Y. M., Roher, A. E., Brachova, L., Shen, Y., Sue, L., Beach, T., Kurth, J. H., Rydel, R. E. & Rogers, J. (1999). *Am J Pathol*, *155*, 853-862
[124] McLean, C. A., Cherny, R. A., Fraser, F. W., Fuller, S. J., Smith, M. J., Beyreuther, K., Bush, A. I. & Masters, C. L. (1999). *Ann Neurol*, *46*, 860-866
[125] Triantafilou, M., Miyake, K., Golenbock, D. T. & Triantafilou, K. (2002). *J Cell Sci*, *115*, 2603-2611
[126] Montixi, C., Langlet, C., Bernard, A.-M., Thimonier, J., Dubois, C., Wurbel, M.-A., Chauvin, J.-P., Pierres, M. & He, H.-T. (1998). *EMBO J*, *17*, 5334-5348
[127] Werlen, G. & Palmer, E. (2002). *Current opinion in immunology*, *14*, 299-305
[128] Dykstra, M. L., Cherukuri, A. & Pierce, S. K. (2001). *Traffic*, *2*, 160-166
[129] Triantafilou, M. & Triantafilou, K. (2002). *Trends in Immunology*, *23*, 301-304
[130] Hunter, T. (2000). *Cell*, *100*, 113-127
[131] Hering, H., Lin, C. C. & Sheng, M. (2003). *J Neurosci*, *23*, 3262-3271
[132] Mauch, D. H., Nagler, K., Schumacher, S., Goritz, C., Muller, E.-C., Otto, A. & Pfrieger, F. W. (2001). *Science*, *294*, 1354-1357
[133] Pfrieger, F. W. (2003). *Biochim Biophys Acta*, *1610*, 271-280
[134] Goritz, C., Mauch, D. H. & Pfrieger, F. W. (2005). *Mol & Cell Neurosci*, *29*, 190-201
[135] Takamori, S., Holt, M., Stenius, K., Lemke, E. A., Grønborg, M., Riedel, D., Urlaub, H., Schenck, S., Brügger, B., Ringler, P., Müller, S. A., Rammner, B., Gräter, F., Hub, J. S., De Groot, B. L., Mieskes, G., Moriyama, Y., Klingauf, J., Grubmüller, H., Heuser, J., Wieland, F. & Jahn, R. (2006). *Cell*, *127*, 831-846
[136] Linetti, A., Fratangeli, A., Taverna, E., Valnegri, P., Francolini, M., Cappello, V., Matteoli, M., Passafaro, M. & Rosa, P. (2010). *J Cell Sci*, *123*, 595-605
[137] Mailman, T., Hariharan, M. & Karten, B. (2011). *J Neurochem*, no-no
[138] Baenziger, J. E., Morris, M. L., Darsaut, T. E. & Ryan, S. E. (2000). *J.Biol.Chem.*, *275*, 777-784

[139] Moskowitz, N., Puszkin, S. & Schook, W. (1983). *J Neurochem, 41*, 1576-1586

[140] Moskowitz, N., Schook, W. & Puszkin, S. (1984). *Brain Res, 290*, 273-279

[141] Moskowitz, N., Schook, W. & Puszkin, S. (1982). *Science, 216*, 305-307

[142] Schaeffer, E. L. & Gattaz, W. F. (2005). *Psychopharmacology (Berl)., 181*, 392-400

[143] Brown, W. J., Chambers, K. & Doody, A. (2003). *Traffic, 4*, 214-221

[144] Chambers, K., Judson, B. & Brown, W. J. (2005). *J Cell Sci, 118*, 3061-3071

[145] Bai, J., Hu, Z., Dittman, J. S., Pym, E. C. G. & Kaplan, J. M. (2010). *Cell, 143*, 430-441

[146] Schmidt, A., Wolde, M., Thiele, C., Fest, W., Kratzin, H., Podtelejnikov, A. V., Witke, W., Huttner, W. B. & Soling, H.-D. (1999). *Nature, 401*, 133-141

[147] Voglmaier, S. M., Kam, K., Yang, H., Fortin, D. L., Hua, Z., Nicoll, R. A. & Edwards, R. H. (2006). *Neuron, 51*, 71-84

[148] Sang, N., Zhang, J., Marcheselli, V., Bazan, N. G. & Chen, C. (2005). *J.Neuroscience, 25*, 9858-9870

[149] Chen, C. & Bazan, N. G. (2005). *J Neurphysiol, 93*, 929-941

[150] Koch, H., Huh, S. E., Elsen, F. P., Carroll, M. S., Hodge, R. D., Bedogni, F., Turner, M. S., Hevner, R. F. & Ramirez, J. M. (2010). *J Neurosci, 30*, 11678-11687

[151] Sun, G. Y., Xu, J., Jensen, M. D. & Simonyi, A. (2004). *J Lipid Res, 45*, 205-213

[152] Farooqui, A. A., Horrocks, L. A. & Farooqui, T. (2007). *J Neurochem, 101*, 577-599

[153] Lukiw, W. J., Cui, J. G., Marcheselli, V. L., Bodker, M., Botkjaer, A., Gotlinger, K., Serhan, C. N. & Bazan, N. G. (2005). *Journal of Clinical Investigation, 115*, 2774-2783

[154] Bazan, N. G. (2005). *Brain Pathol, 15*, 159-166

[155] Francescangeli, E., Lang, D., Dreyfus, H., Boila, A., Freysz, L. & Goracci, G. (1997). *Neurochem Res, 22*, 1299-1307

[156] Francescangeli, E., Domanska-Janik, K. & Goracci, G. (1996). *J Lipid Mediat Cell Signal, 14*, 89-98

[157] Shindou, H., Hishikawa, D., Nakanishi, H., Harayama, T., Ishii, S., Taguchi, R. & Shimizu, T. (2007). *J.Biol.Chem., 282*, 6532-6539

[158] Marcheselli, V. L., Rossowska, M. J., Domingo, M. T., Braquet, P. & Bazan, N. G. (1990). *J Biol Chem., 265*, 9140-9145

[159] Wieraszko, A., Li, G., Kornecki, E., Hogan, M. V. & Ehrlich, Y. H. (1993). *Neuron*, *10*, 553-557

[160] Kato, K., Clark, G. D., Bazan, N. G. & Zorumski, C. F. (1994). *Nature*, *367*, 175-179

[161] Chen, C., Magee, J. C., Marcheselli, V., Hardy, M. & Bazan, N. G. (2001). *J Neurophysiol*, *85*, 384-390

[162] Bazan, N. G. (1998). *Prog Brain Res*, *118*, 281-291

[163] Shelat, P. B., Chalimoniuk, M., Wang, J. H., Strosznajder, J. B., Lee, J. C., Sun, A. Y., Simonyi, A. & Sun, G. Y. (2008). *J Neurochem*, *106*, 45-55

[164] Singh, I. N., Sorrentino, G., Sitar, D. S. & Kanfer, J. N. (1998). *Brain Res*, *800*, 275-281

[165] Hicks, J. B., Lai, Y., Sheng, W., Yang, X., Zhu, D., Sun, G. Y. & Lee, J. C. (2008). *Biochim Biophys Acta*, *1778*, 2512-2519

[166] Desbene, C., Malaplate-Armand, C., Youssef, I., Garcia, P., Stenger, C., Sauvee, M., Fischer, N., Rimet, D., Koziel, V., Escanye, M. C., Oster, T., Kriem, B., Yen, F. T., Pillot, T. & Olivier, J. L. (2012). *Neurobiol Aging*, *33*, 1123 e1117-1129

[167] Sanchez-Mejia, R. O., Newman, J. W., Toh, S., Yu, G.-Q., Zhou, Y., Halabisky, B., Cisse, M., Scearce-Levie, K., Cheng, I. H., Gan, L., Palop, J. J., Bonventre, J. V. & Mucke, L. (2008). *Nat Neurosci*, *11*, 1311-1318

[168] Bate, C., Reid, S. & Williams, A. (2004). *J.Biol.Chem.*, *279*, 36405-36411

[169] Bate, C., Salmona, M. & Williams, A. (2004). *Neuroreport*, *15*, 509-513

[170] Farooqui, A. A., Ong, W.-Y. & Horrocks, L. A. (2006). *Phamacol. Rev*, *58*, 591-620

[171] Farooqui, A. A. & Horrocks, L. A. (2004). *Prostaglandins, Leukotrienes and Essential Fatty Acids*, *71*, 161-169

[172] Williams, A., Van Dam, A. M., Ritchie, D., Eikelenboom, P. & Fraser, H. (1997). *Brain Res.*, *754*, 171-180

[173] Minghetti, L., Cardone, F., Greco, A., Puopolo, M., Levi, G., Green, A. J., Knight, R. & Pocchiari, M. (2002). *Neurology*, *58*, 127-129

[174] Minghetti, L., Greco, A., Cardone, F., Puopolo, M., Ladogana, A., Almonti, S., Cunningham, C., Perry, V. H., Pocchiari, M. & Levi, G. (2000). *J.Neuropathol.Exp.Neurol.*, *59*, 866-871

[175] Montine, T. J., Sidell, K. R., Crews, B. C., Markesbery, W. R., Marnett, L. J., Roberts, L. J. & Morrow, J. D. (1999). *Neurology*, *53*, 1495-1498

[176] Ryan, S. D., Whitehead, S. N., Swayne, L. A., Moffat, T. C., Hou, W., Ethier, M., Bourgeois, A. J. G., Rashidian, J., Blanchard, A. P., Fraser, P. E., Park, D. S., Figeys, D. & Bennett, S. A. L. (2009). *Proc Natl Acad Sci USA*, *106*, 20936-20941

[177] Bellizzi, M. J., Lu, S. M., Masliah, E. & Gelbard, H. A. (2005). *J Clin Invest*, *115*, 3185-3192

[178] Cimino, P. J., Keene, C. D., Breyer, R. M., Montine, K. S. & Montine, T. J. (2008). *Curr Med Chem*, *15*, 1863-1869

[179] Liang, X., Wang, Q., Hand, T., Wu, L., Breyer, R. M., Montine, T. J. & Andreasson, K. (2005). *J. Neurosci.*, *25*, 10180-10187

[180] Kim, H. W., Rapoport, S. I. & Rao, J. S. (2011). *Mol Psychiatry*, *16*, 419-428

[181] Gattaz, W. F., Valente, K. D., Raposo, N. R. B., Vincentiis, S. & Talib, L. L. (2011). *Journal of Psychiatric Research*, *45*, 1617-1620

[182] Bazan, N. G., Tu, B. & Rodriguez de Turco, E. B. (2002). What synaptic lipid signaling tells us about seizure-induced damage and epileptogenesis. in *Progress in Brain Research* (Thomas Sutula, A. P. ed.)., Elsevier. 175-185

[183] Hudson, C., Gotowiec, A., Seeman, M., Warsh, J. & Ross, B. M. (1999). *Biological Psychiatry*, *46*, 401-405

[184] Klivenyi, P., Beal, M. F., Ferrante, R. J. &reassen, O. A., Wermer, M., Chin, M.-R. & Bonventre, J. V. (1998). *J Neurochem*, *71*, 2634-2637

[185] Malgorzata, C., Anna, S., Elzbieta, Z., Adam, S., Josef, L. & Joanna, B. S. (2009). *J Neurochem*, *110*, 307-317

[186] Moore, D. J., Masliah, E., Rippeth, J. D., Gonzalez, R., Carey, C. L., Cherner, M., Ellis, R. J., Achim, C. L., Marcotte, T. D., Heaton, R. K. & Grant, I. (2006). *AIDS*, *20*, 879-887

[187] Zhou, L., Diefenbach, E., Crossett, B., Tran, S., Ng, T., Rizos, H., Rua, R., Wang, B., Kapur, A., Gandhi, K., Brew, B. & Saksena, N. (2010). *Mol Neurodegener*, *5*, 27

[188] Behnisch, T., Francesconi, W. & Sanna, P. P. (2004). *Brain Res*, *1012*, 187-189

[189] Kim, H. J., Martemyanov, K. A. & Thayer, S. A. (2008). *J. Neurosci.*, *28*, 12604-12613

[190] Iskander, S., Walsh, K. & Hammond, R. (2004). *J Neuroinflammation*, *1*, 7

[191] Rao, J. S., Kim, H.-W., Kellom, M., Greenstein, D., Chen, M., Kraft, A. D., Harry, G. J., Rapoport, S. I. & Basselin, M. (2011). *Journal of Neuroinflammation*, *8*, 101

[192] Serradji, N., Martin, M., Bensaid, O., Cisternino, S., Rousselle, C., Dereuddre-Bosquet, N., Huet, J., Redeuilh, C., Lamouri, A., Dong, C. Z., Clayette, P., Scherrmann, J. M., Dormont, D. & Heymans, F. (2004). *J.Med.Chem.*, *47*, 6410-6419

[193] Eggert, D., Dash, P. K., Serradji, N., Dong, C.-Z., Clayette, P., Heymans, F., Dou, H., Gorantla, S., Gelbard, H. A., Poluektova, L. & Gendelman, H. E. (2009). *J Neuroimmunol*, *213*, 47-59

[194] Gelbard, H. A., Nottet, H. S., Swindells, S., Jett, M., Dzenko, K. A., Genis, P., White, R., Wang, L., Choi, Y. B. & Zhang, D. (1994). *J.Virol.*, *68*, 4628-4635

[195] Nalefski, E. A., Sultzman, L. A., Martin, D. M., Kriz, R. W., Towler, P. S., Knopf, J. L. & Clark, J. D. (1994). *J.Biol.Chem.*, *269*, 18239-18249

[196] Evans, J. H. & Leslie, C. C. (2004). *J.Biol.Chem.*, *279*, 6005-6016

[197] Ball, A., Nielsen, R., Gelb, M. H. & Robinson, B. H. (1999). *Proc Natl Acad Sci USA*, *96*, 6637-6642

[198] Akiba, S., Ii, H., Yoneda, Y. & Sato, T. (2004). *Biochim Biophys Acta*, *1686*, 77-84

[199] Gaudreault, S. B., Chabot, C., Gratton, J. P. & Poirier, J. (2004). *J.Biol.Chem.*, *279*, 356-362

[200] Bate, C. & Williams, A. (2011). *Prion*, *5*

[201] Bate, C. & Williams, A. (2011). *J.Biol.Chem.*, *286*, 37955 - 37963

[202] Gong, Y., Chang, L., Viola, K. L., Lacor, P. N., Lambert, M. P., Finch, C. E., Krafft, G. A. & Klein, W. L. (2003). *Proc Natl Acad Sci USA*, *100*, 10417-10422

[203] Lacor, P. N., Buniel, M. C., Chang, L., Fernandez, S. J., Gong, Y., Viola, K. L., Lambert, M. P., Velasco, P. T., Bigio, E. H., Finch, C. E., Krafft, G. A. & Klein, W. L. (2004). *J Neurosci*, *24*, 10191-10200

[204] Renner, M., Lacor, P. N., Velasco, P. T., Xu, J., Contractor, A., Klein, W. L. & Triller, A. (2010). *Neuron*, *66*, 739-754

[205] Lorenzo, A., Yuan, M., Zhang, Z., Paganetti, P. A., Sturchler-Pierrat, C., Staufenbiel, M., Mautino, J., Vigo, F. S., Sommer, B. & Yankner, B. A. (2000). *Nat Neurosci*, *3*, 460-464

[206] Yan, S. D., Chen, X., Fu, J., Chen, M., Zhu, H., Roher, A., Slattery, T., Zhao, L., Nagashima, M., Morser, J., Migheli, A., Nawroth, P., Stern, D. & Schmidt, A. M. (1996). *Nature*, *382*, 685-691

[207] Lauren, J., Gimbel, D. A., Nygaard, H. B., Gilbert, J. W. & Strittmatter, S. M. (2009). *Nature*, *457*, 1128-1132

[208] Sunyach, C., Jen, A., Deng, J., Fitzgerald, K. T., Frobert, Y., Grassi, J., McCaffrey, M. W. & Morris, R. (2003). *EMBO J.*, *22*, 3591-3601

[209] Brown, D. R. (2001). *Trends Neurosci*, *24*, 85-90
[210] Herms, J., Tings, T., Gall, S., Madlung, A., Giese, A., Siebert, H., Schurmann, P., Windl, O., Brose, N. & Kretzschmar, H. (1999). *J Neurosci*, *19*, 8866-8875
[211] Linden, R., Cordeiro, Y. & Lima, L. M. (2012). *Cell Mol Life Sci*, *69*, 1105-1124
[212] Chiesa, R., Piccardo, P., Biasini, E., Ghetti, B. & Harris, D. A. (2008). *J Neurosci*, *28*, 13258-13267
[213] Stuermer, C. A. O., Langhorst, M. F., Wiechers, M. F., Legler, D. F., Hannbeck von Hanwehr, S., Guse, A. H. & Plattner, H. (2004). *FASEB J.*, *14*, 1731-1733
[214] Solforosi, L., Criado, J. R., McGavern, D. B., Wirz, S., Sanchez-Alavez, M., Sugama, S., DeGiorgio, L. A., Volpe, B. T., Wiseman, E., Abalos, G., Masliah, E., Gilden, D., Oldstone, M. B., Conti, B. & Williamson, R. A. (2004). *Science*, *303*, 1514-1516
[215] Cleary, J. P., Walsh, D. M., Hofmeister, J. J., Shankar, G. M., Kuskowski, M. A., Selkoe, D. J. & Ashe, K. H. (2005). *Nat Neurosci*, *8*, 79-84
[216] Giuffrida, M. L., Caraci, F., Pignataro, B., Cataldo, S., De Bona, P., Bruno, V., Molinaro, G., Pappalardo, G., Messina, A., Palmigiano, A., Garozzo, D., Nicoletti, F., Rizzarelli, E. & Copani, A. (2009). *J Neurosci*, *29*, 10582-10587
[217] Qing, H., He, G., Ly, P. T. T., Fox, C. J., Staufenbiel, M., Cai, F., Zhang, Z., Wei, S., Sun, X., Chen, C.-H., Zhou, W., Wang, K. & Song, W. (2008). *J Exp Med*, *205*, 2781-2789
[218] Dou, H., Birusingh, K., Faraci, J., Gorantla, S., Poluektova, L. Y., Maggirwar, S. B., Dewhurst, S., Gelbard, H. A. & Gendelman, H. E. (2003). *J Neurosci*, *23*, 9162-9170
[219] Monti, B., Gatta, V., Piretti, F., Raffaelli, S. S., Virgili, M. & Contestabile, A. (2010). *Neurotox Res*, *17*, 130-141
[220] Chang, M. C. J., Contreras, M. A., Rosenberger, T. A., Rintala, J. J. O., Bell, J. M. & Rapoport, S. I. (2001). *J Neurochem*, *77*, 796-803
[221] Bate, C., Tayebi, M. & Williams, A. (2008). *Mol Neurodegener*, *3*, 1
[222] Ahlemeyer, B. & Krieglstein, J. (2003). *Pharmacopsychiatry* 36 Suppl 1:, S8-14
[223] Bastianetto, S. & Quirion, R. (2002). *Neurobiol Aging*, *23*, 891-897
[224] Le Bars, P. L., Velasco, F. M., Ferguson, J. M., Dessain, E. C., Kieser, M. & Hoerr, R. (2002). *Neuropsychobiology*, *45*, 19-26

[225] De Felice, F. G., Vieira, M. N., Bomfim, T. R., Decker, H., Velasco, P. T., Lambert, M. P., Viola, K. L., Zhao, W. Q., Ferreira, S. T. & Klein, W. L. (2009). *Proc Natl Acad Sci USA*, *106*, 1971-1976

[226] Reger, M. A., Watson, G. S., Green, P. S., Wilkinson, C. W., Baker, L. D., Cholerton, B., Fishel, M. A., Plymate, S. R., Breitner, J. C. S., DeGroodt, W., Mehta, P. & Craft, S. (2008). *Neurology*, *70*, 440-448

[227] Wilson, R., Bate, C., Boshuizen, R., Williams, A. & Brewer, J. (2007). *BMC Neuroscience*, *8*, 99

[228] Bate, C., Salmona, M., Diomede, L. & Williams, A. (2004). *J.Biol.Chem.*, *279*, 14983-14990

[229] Hagiwara, K., Nakamura, Y., Nishijima, M. & Yamakawa, Y. (2007). *Biol Pharm Bull*, *30*, 835-838

[230] Bate, C., Salmona, M., Diomede, L. & Williams, A. (2004). *J.Biol.Chem.*, *279*, 14983-14990

[231] Allen, J. A., Halverson-Tamboli, R. A. & Rasenick, M. M. (2007). *Nat Rev Neurosci*, *8*, 128-140

[232] Wilson, B. S., Steinberg, S. L., Liederman, K., Pfeiffer, J. R., Surviladze, Z., Zhang, J., Samelson, L. E., Yang, L. h., Kotula, P. G. & Oliver, J. M. (2004). *Mol Biol Cell*, *15*, 2580-2592

[233] Movahedi, S. & Hooper, N. M. (1997). *Biochem J*, *326* (Pt 2)., 531-537

[234] Müller, G., Jung, C., Wied, S., Welte, S. & Frick, W. (2001). *Biochemistry*, *40*, 14603-14620

[235] Müller, G., Dearey, E. A. & Punter, J. (1993). *Biochem J*, *289* (Pt 2)., 509-521

[236] Bate, C., Tayebi, M., Diomede, L., Salmona, M. & Williams, A. (2009). *PLoS ONE*, *4*, e8221

[237] Taylor, D. R. & Hooper, N. M. (2006). *Mol Membr Biol*, *23*, 89-99

[238] Chatterjee, S. & Mayor, S. (2001). *Cell Mol Life Sci*, *58*, 1969-1987

[239] Major, S. & Reizman, H. (2004). *Nature reviews (Molecular Cell Biology).*, *5*, 110 -120

[240] Ikezawa, H. (2002). *Biol Pharm Bull*, *25*, 409-417

[241] Anderson, R. G. W. & Jacobson, K. (2002). *Science*, *296*, 1821-1825

[242] Stahl, N., Baldwin, M. A., Hecker, R., Pan, K. M., Burlingame, A. L. & Prusiner, S. B. (1992). *Biochemistry*, *31*, 5043-5053

[243] Homans, S. W., Ferguson, M. A. J., Dwek, R. A., Rademacher, T. W., Anand, R. & Williams, A. F. (1988). *Nature*, *333*, 269-272

[244] Madore, N., Smith, K. L., Graham, C. H., Jen, A., Brady, K., Hall, S. & Morris, R. (1999). *EMBO J.*, *18*, 6917-6926

[245] Brugger, B., Graham, C., Leibrecht, I., Mombelli, E., Jen, A., Wieland, F. & Morris, R. (2004). *J.Biol.Chem.*, *279*, 7530-7536

[246] Nicholson, T. B. & Stanners, C. P. (2006). *J Cell Biol*, *175*, 647-659

[247] Premkumar, D. R., Fukuoka, Y., Sevlever, D., Brunschwig, E., Rosenberry, T. L., Tykocinski, M. L. & Medof, M. E. (2001). *J Cell Biochem*, *82*, 234-245

[248] Medof, M. E., Nagarajan, S. & Tykocinski, M. L. (1996). *FASEB J.*, *10*, 574-586

[249] Bate, C., Tayebi, M. & Williams, A. (2010). *J.Biol.Chem.*, *285*, 22017 - 22026

[250] Chesebro, B., Trifilo, M., Race, R., Meade-White, K., Teng, C., LaCasse, R., Raymond, L., Favara, C., Baron, G., Priola, S., Caughey, B., Masliah, E. & Oldstone, M. (2005). *Science*, *308*, 1435-1439

[251] Chatterjee, S., Smith, E. R., Hanada, K., Stevens, V. L. & Mayor, S. (2001). *EMBO J*, *20*, 1583-1592

[252] Marijanovic, Z., Caputo, A., Campana, V. & Zurzolo, C. (2009). *PLoS Pathog*, *5*, e1000426

[253] Caetano, F. A., Beraldo, F. H., Hajj, G. N. M., Guimaraes, A. L., Jürgensen, S., Wasilewska-Sampaio, A. P., Hirata, P. H. F., Souza, I., Machado, C. F., Wong, D. Y. L., De Felice, F. G., Ferreira, S. T., Prado, V. F., Rylett, R. J., Martins, V. R. & Prado, M. A. M. (2011). *J Neurochem*, *117*, 538-553

[254] Bryleva, E. Y., Rogers, M. A., Chang, C. C., Buen, F., Harris, B. T., Rousselet, E., Seidah, N. G., Oddo, S., LaFerla, F. M., Spencer, T. A., Hickey, W. F. & Chang, T. Y. (2010). *Proc Natl Acad Sci U S A*, *107*, 3081-3086

[255] Bate, C., Tayebi, M. & Williams, A. (2008). *Neuropharmacology*, *54*, 1247-1253

[256] Nishio, H., Takeuchi, T., Hata, F. & Yagasaki, O. (1996). *Biochem J*, *318* (Pt 3)., 981-987

[257] van Meer, G. & Vaz, W. L. (2005). *EMBO Rep*, *6*, 418-419

[258] Huttner, W. B. & Schmidt, A. (2000). *Curr Opin Neurobiol*, *10*, 543-551

[259] Kooijman, E. E., Chupin, V., De Kruijff, B. & Burger, K. N. J. (2003). *Traffic*, *4*, 162-174

[260] Hishikawa, D., Shindou, H., Kobayashi, S., Nakanishi, H., Taguchi, R. & Shimizu, T. (2008). *Proc Natl Acad Sci U S A*, *105*, 2830-2835

[261] Farooqui, A. A., Horrocks, L. A. & Farooqui, T. (2000). *J Mol Neurosci.*, *14*, 123-135

[262] Song, W. & Zinsmaier, K. E. (2003). *Neuron*, *40*, 665-667

[263] Weigert, R., Silletta, M. G., Spano, S., Turacchio, G., Cericola, C., Colanzi, A., Senatore, S., Mancini, R., Polishchuk, E. V., Salmona, M., Facchiano, F., Burger, K. N. J., Mironov, A., Luini, A. & Corda, D. (1999). *Nature*, *402*, 429-433

[264] Carrasco, M. P., Jimenez-Lopez, J. M., Rios-Marco, P., Segovia, J. L. & Marco, C. (2010). *Br J Pharmacol*, *160*, 355-366

[265] Jimenez-Lopez, J. M., Rios-Marco, P., Marco, C., Segovia, J. L. & Carrasco, M. P. (2010). *Lipids Health Dis*, *9*, 33

[266] Ausili, A., Torrecillas, A., Aranda, F. J., Mollinedo, F., Gajate, C., Corbalán-García, S., de Godos, A. & Gómez-Fernández, J. C. (2008). *The Journal of Physical Chemistry B*, *112*, 11643-11654

[267] van der Luit, A. H., Budde, M., Ruurs, P., Verheij, M. & van Blitterswijk, W. J. (2002). *J.Biol.Chem.*, *277*, 39541-39547

[268] Drecktrah, D., Chambers, K., Racoosin, E. L., Cluett, E. B., Gucwa, A., Jackson, B. & Brown, W. J. (2003). *Molecular biology of the cell*, *14*, 3459-3469

[269] Drecktrah, D., Chambers, K., Racoosin, E. L., Cluett, E. B., Gucwa, A., Jackson, B. & Brown, W. J. (2003). *Mol Biol Cell*, *14*, 3459-3469

[270] Kobayashi, T., Beuchat, M.-H., Lindsay, M., Frias, S., Palmiter, R. D., Sakuraba, H., Parton, R. G. & Gruenberg, J. (1999). *Nat Cell Biol*, *1*, 113-118

[271] Chevallier, J., Chamoun, Z., Jiang, G., Prestwich, G., Sakai, N., Matile, S., Parton, R. G. & Gruenberg, J. (2008). *J.Biol.Chem.* 283, 27871-27880

[272] Stampfer, M. J., Kang, J. H., Chen, J., Cherry, R. & Grodstein, F. (2005). *N Engl J Med*, *352*, 245-253

[273] Mukamal, K. J., Kuller, L. H., Fitzpatrick, A. L., Longstreth, W. T., Jr., Mittleman, M. A. & Siscovick, D. S. (2003). *JAMA*, *289*, 1405-1413

[274] Collins, M. A., Neafsey, E. J., Mukamal, K. J., Gray, M. O., Parks, D. A., Das, D. K. & Korthuis, R. J. (2009). *Alcohol Clin Exp Res*, *33*, 206-219

[275] Bate, C. & Williams, A. (2011). *Neuropharmacology*, *61*, 1406 - 1412

[276] Saito, M., Saito, M., Cooper, T. B. & Vadasz, C. (2005). *Alcohol Clin Exp Res*, *29*, 1374-1383

[277] Rosso, A., Mossey, J. & Lippa, C. F. (2008). *Am J Alzheimers Dis Other Demen*, *23*, 417-422

[278] Eskelinen, M. H., Ngandu, T., Tuomilehto, J., Soininen, H. & Kivipelto, M. (2009). *J Alzheimers Dis*, *16*, 85-91

[279] Ascherio, A., Zhang, S. M., Hernan, M. A., Kawachi, I., Colditz, G. A., Speizer, F. E. & Willett, W. C. (2001). *Ann Neurol*, *50*, 56-63
[280] Ross, G. W., Abbott, R. D., Petrovitch, H., Morens, D. M., Grandinetti, A., Tung, K. H., Tanner, C. M., Masaki, K. H., Blanchette, P. L., Curb, J. D., Popper, J. S. & White, L. R. (2000). *JAMA*, *283*, 2674-2679
[281] Ritchie, K., Carriere, I., de Mendonca, A., Portet, F., Dartigues, J. F., Rouaud, O., Barberger-Gateau, P. & Ancelin, M. L. (2007). *Neurology*, *69*, 536-545
[282] de Lau, L. M. L., Koudstaal, P. J., Witteman, J. C. M., Hofman, A. & Breteler, M. M. B. (2006). *Neurology*, *67*, 315-318
[283] Aisen, P. S., Schneider, L. S., Sano, M., Diaz-Arrastia, R., van Dyck, C. H., Weiner, M. F., Bottiglieri, T., Jin, S., Stokes, K. T., Thomas, R. G., Thal, L. J. & for the Alzheimer Disease Cooperative Study. (2008). *JAMA*, *300*, 1774-1783
[284] Sung, S., Yao, Y., Uryu, K., Yang, H., Lee, V. M.-Y., Trojanowski, J. Q. & Pratico, D. (2003). *FASEB J.*, 03-0961fje
[285] Khanna, S., Roy, S., Slivka, A., Craft, T. K., Chaki, S., Rink, C., Notestine, M. A., DeVries, A. C., Parinandi, N. L. & Sen, C. K. (2005). *Stroke*, *36*, 2258-2264
[286] Politis, A., Olgiati, P., Malitas, P., Albani, D., Signorini, A., Polito, L., De Mauro, S., Zisaki, A., Piperi, C., Stamouli, E., Mailis, A., Batelli, S., Forloni, G., De Ronchi, D., Kalofoutis, A., Liappas, I. & Serretti, A. (2010). *J Alzheimers Dis*, *19*, 481-488
[287] Chan, A., Tchantchou, F., Graves, V., Rozen, R. & Shea, T. B. (2008). *J Nutr Health Aging*, *12*, 252-261
[288] Luchsinger, J. A. & Mayeux, R. (2004). *Lancet Neurol*, *3*, 579-587
[289] Chandra, V., Jasti, J., Kaur, P., Betzel, C., Srinivasan, A. & Singh, T. P. (2002). *J Mol Biol*, *320*, 215-222
[290] Khanna, S., Parinandi, N. L., Kotha, S. R., Roy, S., Rink, C., Bibus, D. & Sen, C. K. (2010). *J Neurochem*, *112*, 1249-1260
[291] Sen, C. K., Rink, C. & Khanna, S. (2010). *J Am Coll Nutr*, *29*, 314S-323S
[292] Pentland, A. P., Morrison, A. R., Jacobs, S. C., Hruza, L. L., Hebert, J. S. & Packer, L. (1992). *Journal of Biological Chemistry*, *267*, 15578-15584
[293] Sen, C. K., Khanna, S. & Roy, S. (2004). *Ann N Y Acad Sci*, *1031*, 127-142
[294] Chi, E. Y., Frey, S. L. & Lee, K. Y. C. (2007). *Biochemistry*, *46*, 1913-1924

[295] Ariga, T., McDonald, M. P. & Yu, R. K. (2008). *J Lipid Res*, *49*, 1157-1175

[296] Naslavsky, N., Shmeeda, H., Friedlander, G., Yanai, A., Futerman, A. H., Barenholz, Y. & Taraboulos, A. (1999). *J.Biol.Chem.*, *274*, 20763-20771

[297] Fyrst, H. & Saba, J. D. (2010). *Nat Chem Biol*, *6*, 489-497

[298] Gulbins, E. & Grassme, H. (2002). *Biochim Biophys Acta*, *1585*, 139-145

[299] Bollinger, C. R., Teichgraber, V. & Gulbins, E. (2005). *Biochim Biophys Acta*, *1746*, 284-294

[300] Colombaioni, L. & Garcia-Gil, M. (2004). *Brain Res Rev*, *46*, 328-355

[301] Staneva, G., Momchilova, A., Wolf, C., Quinn, P. J. & Koumanov, K. (2009). *Biochimica et Biophysica Acta (BBA). - Biomembranes*, *1788*, 666-675

[302] van Blitterswijk, W. J., van der Luit, A. H., Veldman, R. J., Verheij, M. & Borst, J. (2003). *Biochem. J.*, *369*, 199-211

[303] Latorre, E., Aragones, M. D., Fernandez, I. & Catalan, R. E. (1999). *Eur J Biochem*, *262*, 308-314

[304] Nakamura, H., Wakita, S., Suganami, A., Tamura, Y., Hanada, K. & Murayama, T. (2010). *J Lipid Res*, *51*, 720-728

[305] Lamour, N. F., Subramanian, P., Wijesinghe, D. S., Stahelin, R. V., Bonventre, J. V. & Chalfant, C. E. (2009). *J.Biol.Chem.*, *284*, 26897-26907

[306] He, X., Huang, Y., Li, B., Gong, C. X. & Schuchman, E. H. (2008). *Neurobiol Aging*

[307] Schmalzbauer, R., Eigenbrod, S., Winoto-Morbach, S., Xiang, W., Schutze, S., Bertsch, U. & Kretzschmar, H. A. (2008). *J Neurochem*, *106*, 1459-1470

[308] Cutler, R. G., Kelly, J., Storie, K., Pedersen, W. A., Tammara, A., Hatanpaa, K., Troncoso, J. C. & Mattson, M. P. (2004). *Proc Natl Acad Sci U S A*, *101*, 2070-2075

[309] Liu, J.-P., Tang, Y., Zhou, S., Toh, B. H., McLean, C. & Li, H. (2010). *Mol Cell Neurosci*, *43*, 33-42

[310] Solfrizzi, V., Colacicco, A. M., D'Introno, A., Capurso, C., Torres, F., Rizzo, C., Capurso, A. & Panza, F. (2006). *Neurobiol Aging*, *27*, 1694-1704

[311] Bate, C., Tayebi, M., Salmona, M., Diomede, L. & Williams, A. (2010). *Neurotox Res*, *17*, 203-214

[312] Zacharias, D. A., Violin, J. D., Newton, A. C. & Tsien, R. Y. (2002). *Science*, *296*, 913-916
[313] Morris, R. J., Parkyn, C. J. & Jen, A. (2006). *FEBS Letters*, *580*, 5565-5571
[314] Shyng, S. L., Huber, M. T. & Harris, D. A. (1993). *J.Biol.Chem.*, *268*, 15922-15928
[315] Bate, C. & Williams, A. (2011). *J.Biol.Chem.*, *286*, 8752 - 8758
[316] Bueler, H., Aguzzi, A., Sailer, A., Greiner, R. A., Autenried, P., Aguet, M. & Weissmann, C. (1993). *Cell*, *73*, 1339-1347
[317] Mallucci, G., Dickinson, A., Linehan, J., Klohn, P. C., Brandner, S. & Collinge, J. (2003). *Science*, *302*, 871-874
[318] Beranger, F., Mange, A., Goud, B. & Lehmann, S. (2002). *J.Biol.Chem.*, *277*, 38972-38977
[319] Nichols, B. J., Kenworthy, A. K., Polishchuk, R. S., Lodge, R., Roberts, T. H., Hirschberg, K., Phair, R. D. & Lippincott-Schwartz, J. (2001). *J Cell Biol*, *153*, 529-541
[320] Gouras, G. K., Tampellini, D., Takahashi, R. H. & Capetillo-Zarate, E. (2010). *Acta Neuropathol*, *119*, 523-541
[321] Takahashi, R. H., Milner, T. A., Li, F., Nam, E. E., Edgar, M. A., Yamaguchi, H., Beal, M. F., Xu, H., Greengard, P. & Gouras, G. K. (2002). *Am J Path*, *161*, 1869-1879
[322] Salmona, M., Morbin, M., Massignan, T., Colombo, L., Mazzoleni, G., Capobianco, R., Diomede, L., Thaler, F., Mollica, L., Musco, G., Kourie, J. J., Bugiani, O., Sharma, D., Inouye, H., Kirschner, D. A., Forloni, G. & Tagliavini, F. (2003). *J.Biol.Chem.*, *278*, 48146-48153
[323] Iwata, N., Tsubuki, S., Takaki, Y., Shirotani, K., Lu, B., Gerard, N. P., Gerard, C., Hama, E., Lee, H.-J. & Saido, T. C. (2001). *Science*, *292*, 1550-1552
[324] Miller, B. C., Eckman, E. A., Sambamurti, K., Dobbs, N., Chow, K. M., Eckman, C. B., Hersh, L. B. & Thiele, D. L. (2003). *Proc Natl Acad Sci U S A*, *100*, 6221-6226
[325] Sun, B., Zhou, Y., Halabisky, B., Lo, I., Cho, S. H., Mueller-Steiner, S., Devidze, N., Wang, X., Grubb, A. & Gan, L. (2008). *Neuron*, *60*, 247-257
[326] Yamin, R., Zhao, C., O'Connor, P., McKee, A. & Abraham, C. (2009). *Mol Neurodegener*, *4*, 33
[327] Nabi, I. R. & Le, P. U. (2003). *J.Cell Biol.*, *161*, 673-677
[328] Nichols, B. (2003). *J Cell Sci*, *116*, 4707-4714

[329] Thomsen, P., Roepstorff, K., Stahlhut, M. & van Deurs, B. (2002)., *13*, 238-250

[330] Bate, C., Ingham, V. & Williams, A. (2010). *Neuropharmacology*, *60*, 365 - 372

[331] Baxter, A., Fitzgerald, B. J., Hutson, J. L., McCarthy, A. D., Motteram, J. M., Ross, B. C., Sapra, M., Snowden, M. A., Watson, N. S. & Williams, R. J. (1992). *J.Biol.Chem.*, *267*, 11705-11708

[332] McLaughlin, N. J. D., Banerjee, A., Kelher, M. R., Gamboni-Robertson, F., Hamiel, C., Sheppard, F. R., Moore, E. E. & Silliman, C. C. (2006). *J Immunol*, *176*, 7039-7050

[333] McLaughlin, N. J. D., Banerjee, A., Khan, S. Y., Lieber, J. L., Kelher, M. R., Gamboni-Robertson, F., Sheppard, F. R., Moore, E. E., Mierau, G. W., Elzi, D. J. & Silliman, C. C. (2008). *J Immunol*, *180*, 8192-8203

[334] Schulze, H., Kolter, T. & Sandhoff, K. (2009). *Biochim Biophy Acta*, *1793*, 674-683

In: Lipid Rafts
Editor: Dan Sillence

ISBN: 978-1-62948-961-2

Chapter VIII

Neuronal Lipid Raft Perturbations as Key Elements of Early Neurodegeneration: A Challenge in the Search of New Tools for Early Diagnosis

Raquel Marin[1*] and Mario Diaz[2]

[1] Laboratory of Cellular Neurobiology, Department of Physiology, School of Medicine, La Laguna University, Tenerife, Spain

[2] Laboratory of Membrane Physiology and Biophysics, Department of Animal Biology, Faculty of Biology, La Laguna University, Tenerife, Spain

Abstract

Alzheimer's disease (AD) and Parkinson's disease (PD) are the most common neurodegenerative diseases. Indeed, the incidence of these pathologies is increasing, in parallel with growing life expectancy. Diagnosis of AD and PD at early stages is difficult, due in part to the lack of accurate tools when cardinal symptoms are not conclusive. Nevertheless, early diagnosis may help to develop more effective

[*] Corresponding author: email: rmarin@ull.es.

treatments in mild stages when brain damage is still not irreversible. In recent years, a promising field of research has focused on the development of novel biomarkers detectable by non-invasive methods, although the identification of at-risk subjects before the first symptoms arise is still missing. Based upon our recent data, we propose that the profound molecular alterations observed in neuronal lipid rafts at early stages of AD and PD may provide promising candidates as early biomarkers of these diseases. Lipid raft anomalies are related to the modification of lipid composition, which correlates with alterations in the physicochemical properties of these microstructures, as well as protein clustering and protein-lipid interactions. This phenomenon may importantly affect physiological functioning of proteins integrated in signaling platforms (or signalosomes) within lipid rafts, thereby leading to altered (and even toxic) protein signaling that may largely contribute to neurodegeneration.

In this mini-review, we suggest that lipid and protein alterations occurring in lipid rafts during asymptomatic preclinical stages of AD and PD may be reflected in cerebrospinal fluids. At this respect, emerging data are promising, although still require further exploration and validation. If successful, accurate detection and quantification of aberrant levels of lipid raft-related lipid and/or protein products in cerebrospinal fluids of patients at preclinical stages may importantly contribute to the identification of novel therapeutic targets, as well as the development of more specific and efficient diagnosis.

Keywords: Lipid rafts, Alzheimer's disease, Parkinson's disease, signalosomes, cerebrospinal fluid biomarkers

Introduction

Lipid rafts have a particular lipid composition that enable close lipid packing and restricted lateral movement of resident molecules. The lipid signature of lipid rafts involves higher contents of cholesterol, sphingolipids and saturated fatty acids than the bulk plasma membrane. These distinct membrane structures functionally serve to compartmentalize cellular processes by the selective recruitment of certain proteins and lipids within the same microenvironment.

Therefore, they are considered highly dynamic membrane microdomains into which specific subsets of proteins and lipids partition, creating cell-signaling platforms that are essential for neuronal biology [1,2]. It has been proposed that these structures are membrane domains in which neurotransmitter signalling might occur through a clustering of receptors and components of receptor-activated signalling cascades [1]. In the brain, lipid rafts are crucial for learning and memory processes.

Thus, a variety of studies have demonstrated the important role of these microstructures in numerous neuronal activities related to development, synaptic plasticity, communication and preservation, such as cell adhesion, axonal growth and guidance, signal transduction, synaptic transmission and maintenance of dendritic spines [1,3,4].

Lipid rafts are critical players in the progression of neurodegeneration. Thus, abnormal organization of lipid rafts has been observed in different neurodegenerative diseases, including Alzheimer's disease (AD), Parkinson's disease (PD), Huntington's disease and Prion diseases [5,6]. The derangement of these microdomains appears to be the result, at least in part, of aberrant distribution and proportions of representative lipids that comprise their structure, including sphingolipids, cholesterol, polyunsaturated and saturated fatty acids, thereby affecting lipid raft physicochemical properties, which originate from metabolic disturbances of lipid homeostasis during the development of these diseases [7].

Lipid alterations consequently constraint the clustering and activities of specific signaling proteins within signaling platforms (or signalosomes) associated within these domains, a phenomenon that importantly contribute to the development of neuronal aberrant (and even neurotoxic) responses. Increasing data support the idea that derangement of the correct stoichiometry of signalosomes, as part of lipid raft structures, is at the basis of the etiology of AD and PD [8, 9]. Taking into account that proper interactions of raft signaling proteins relies on the correct lipid structure, it is conceivable that changes in, both, lipid and protein components of lipid rafts may underpin the first stages of neuronal impairments in different brain areas, ultimately leading to neurodegenerative diseases [10, 11].

In this order of ideas, as discussed below, the potential detection of early alterations in these microstructures may predict future progression of neurodegeneration. It is therefore conceivable that measurement of altered molecular markers in lipid rafts at preclinical stages of AD and PD may be a promising future tool of early diagnosis of these diseases. However, this hypothesis requires further investigation.

Lipid Raft Impairments in AD and PD

Links between Lipid Rafts Lipids and AD and PD Pathologies

Alzheimer's disease is the most prevalent form of dementia, affecting approximately 1% individuals between 65 and 69 years and 25-50% individuals over 85 years of age. This disease is characterized by a loss of short-term memory and cognitive impairments, as a consequence of progressive neurodegeneration due mainly to two identified pathological traits: 1) accumulation of of amyloid beta (Aβ) peptide, a product of amyloid precursor protein (APP) processing, that forms senile plaques; and 2) aggregation of intracellular protein Tau as a consequence of aberrant hyperphosphorylation, that form neurofibrillary tangles [12, 13].

Although the origin and progression of Alzheimer's disease pathology is not well understood, it is acepted that the formation of senile plaques, as a result of Aβ aggregation, is one of the main hallmarks leading to cognitive decline and memory loss. Aβ formation consists on the sequential cleavage of full-length amyloid precursor protein (APP) by β-secretase (BACE1) and γ-secretase complex. The amyloidogenic processing results in the release of Aβ peptides of between 39–42 residues in length, depending upon the precise cleavage site.

Even though the majority of full-length APP is localized to non-raft regions of the plasma membrane [14], where non-amyloidogenic cleavage by α-secretases precludes Aβ formation [15], there is also consensus that lipid rafts are critical players in the development of AD. Indeed, a pool of amyloid precursor protein is tightly associated with lipid rafts, and its intracellular domain interacts with raft-integrated flotillin-1 to initiate its amyloigenic processing [16, 17]. Moreover, a subset of BACE1 and γ-secretase components (both secretases participating in the cleavage of APP to produce Aβ) partition into lipid rafts, suggesting that Aβ formation occurs in these structures [18, 19]. Both, BACE1 and γ-secretase undergo post-translational S-palmitoylation which aids their targeting to lipid raft domains [20] and γ-secretase complex is regulated by lipid raft scaffolding caveolin-1 [21]. It is also known that lipid rafts affect APP processing not only through favoring interactions between APP and BACE-1, but also by promoting endocytosis of APP, which is required to initiate the release of the neurotoxic β-amyloid peptide [16, 17]. Other authors have suggested the exclusion of non-amyloidogenic α-secretase from lipid rafts, rather than enhancement of lipid raft-localization of BACE and γ-secretase [22]. Moreover, syntaxin-1 participates in lipid raft APP processing through facilitating APP switching between domains [23]. These evidences indicate a dynamic distribution of APP and secretases in the neuronal membrane, with the involvement of lipid rafts. Furthermore, it has been suggested that activation of APP and BACE interaction in lipid rafts involves the participation of the proper molecular microenvironment [11, 14, 24, 25]. These data suggest the presence of a dynamic system leading to Aβ generation with the participation of different lipid raft modulators that require further study.

Besides promoting Aβ formation through APP processing regulation, two other roles are attributed to neuronal lipid rafts in AD: 1) to facilitate Aβ aggregation forming toxic oligomers; and 2) to host specific neuronal receptors through which AD-related neurotoxicity mechanisms and memory impairments are transduced. Aβ can bind to specific lipid classes abundant in lipid rafts, such as gangliosides,which induce structural transitions in the peptide [26]. Thus, Aβ aggregation is enhanced by ganglioside clusters in the presence of cholesterol [27], and ganglioside GM-1 binding to nontoxic Aβ promotes peptide toxic seeding, thereby forming a membrane-active form of Aβ [28, 29]. Furthermore, other evidence suggests that Aβ oligomers may exert deleterious effects through binding to, and causing, the aberrant clustering of lipid raft proteins, including prion protein and glutamate receptors [2]. Even more dramatic is the fact that the interaction of Aβ eptides with raft membranes stimulates the amyloidogenic processing of APP by reducing membrane fluidity and complexing with GM-1 ganglioside, thereby starting a vicious cycle where endogenous Aβ stimulates its own production [30]. The formation of these pathogenic lipid raft-based platforms may be critical for toxic signaling mechanisms that underlie synaptic dysfunction and neuropathology. In support of this, lipid rafts are abundant in mature hippocampal neurons, thus emphasizing their physiological importance within the memory centre of the healthy brain, and may explain why hippocampal neurons are a primary target for amyloid beta (Aβ) oligomer toxicity and destruction in AD [31].

The relationship between membrane lipids and Aβ appears to be bidirectional, since not only the peptide can alter the function of neuronal membrane but also membrane lipids modulate the production and dynamics of Aβ [32]. In this sense, it is conceivable that changes in lipid composition of lipid rafts may largely contribute to AD pathology. Indeed, a positive correlation has been proposed between GM1 and GM2 ganglioside classes, cholesterol loss and formation of senile plaques, as observed in brains from AD patients as well as in mouse models of AD [14, 33-35]. Aβ peptides can also function as as regulators of lipid homeostasis at the cell membrane. Thus, it has been demonstrated that accumulation of Aβ in neuronal membranes leads to sphingomyelinase (SMase) activation and sphingomyelin (SM) depletion, which, in turn, affects cellular trafficking and abnormal APP processing [36]. Moreover, excessive hydrolysis of SM provokes the increase in ceramide levels, which would activate proapoptotic responses [37, 38]. But in the context of the present review, the most impacting effect of augmented SMase-mediated SM cleavage is the disruption of SM-rich platforms, namely lipid rafts, in terms of stability and protein-lipid interactions.

From a biophysical point of view, cholesterol is the lipid molecule mostly responsible for the liquid-ordered state of lipid rafts [25, 26, 39]. Its relationship with Alzheimer's disease has been extensively documented [40, 41]. Although results are conflicting, there appear to exist consensus on the fact that cholesterol affects APP processing in a complex manner, and that a direct relationship exists between higher cholesterol levels and amyloidogenesis, both, in vitro and in vivo [40, 41]. In this context, membrane cholesterol content has an inverse relationship with the membrane-perturbing effects of Aβ oligomers. Accordingly, increased cholesterol do elevate lipid raft abundance, which would favors Aβ formation and, on the contrary, low cholesterol levels will lead to up-regulation of the activity of the α-secretase, ADAM10, and non-amyloidogenic processing of APP [11, 18, 42, 43, 44]. In agreement with this, some reports have shown cholesterol levels to be elevated in AD patients as well as the levels of cholesterol precursors in the mevalonate pathway [11, 45]. In any case, it seems now clear that global cholesterol levels are not necessarily reflective of the status and distribution of lipid rafts [11, 46]. In fact, Analyses of cholesterol contents in lipid rafts from *post mortem* brains (at least frontal cortex), have shown that irrespective of gross brain cholesterol contents, cholesterol levels remained nearly constant during aging and were unaffected in late stages of AD [46].

Other lipids that are acquiring increasing importance in neuronal membrane functionality are long chain polyunsaturated fatty acids (LCPUFA). These fatty acids are key structural elements in brain membranes, and are mostly accumulated in phospholipids outside lipid rafts [47, 48]. Significant reductions of docosahexaenoic acid (DHA) have been shown in different regions of AD postmortem brains, including frontal grey matter and hippocampus [49]. The relevance of this DHA depletion in neuronal phospholipids is enormous since human brain not only is the body organ containing the largest amount of DHA, but also it has a very low capacity to synthesize n-3 PUFA from its precursor α-linoleic acid [50].

Recent evidence have pointed out that beyond its role in whole neuronal membranes, LCPUFA play are critical for lipid raft organization, distribution and stability. Furthermore, a detrimental reduction of LCPUFA levels, in particular DHA, is observed related to AD pathology in human brains and animal models [26, 46, 51]. The concomitant reduction in monoenes observed in the same lipid rafts led to a dramatic reduction in the unsaturation index. One of the physiological consequences of the reduction in monoenes and n-3 LCPUFA in lipid rafts is that it substantially alters the physicochemical properties of these microdomains. Thus, LCPUFA (and specifically DHA) decrease in AD correlates with increased liquid order and viscosity [26]. Consequently, these changes may have profound impact on protein-lipid and protein-protein interactions, as well as the activity of raft proteins present in signalosomes, as discussed below.

Parkinson's disease is mainly characterized by motor impairment and represents, together with AD, the most common neurodegenerative disorder [52]. This disease is characterized by a progressive loss of dopaminergic neurons. Subjects in advanced stages of PD show muscle rigidity, tremor and bradykinesia, mental decline, visual hallucinations and parkinsonism. At the anatomopathological level, one of the main hallmarks is the presence of intraneuronal inclusions (or Lewy bodies) and aberrant neurites showing aberrant aggregates. These formations are constituted mainly by the neuronal protein α-synuclein (α-Syn). The abnormal modifications of this protein (i.e. phosphorylation, nitrosylation and oxidation) are key to the conversion of normally soluble α-Syn into insoluble intracellular aggregates, thus promoting fibril formation and synucleinopathy [53].

Although less studied than AD, there is also some agreement that molecular damage in PD is not limited to specific proteins, but also to levels of fatty acids and lipid classes in different brain areas of PD brains, including basal ganglia and frontal cortex. in the frontal cortex in PD compared with age-matched controls [55-57]. Thus, findings from *post mortem* brains in levodopa-treated PD patients and parkinsonian monkeys have shown abnormally high levels of n-6 LCPUFAs and reduced n-3 LCPUFAs (leading to aberrant n-6/n-3 ratio) compared to controls in the nigrostriatal system, suggesting a role of lipoperoxidative stress in dopaminergic degeneration [54, 55]. Lipid changes are also present in the frontal cortex of PD brains, but strikingly, analysis of fatty acids profiles in PD and incidental PD (iPD, considered an early asymptomatic stage of PD neuropathology) subjects, revealed important increases in the contents of both n-3 and n-6 LCPUFA, mainly DHA and AA, as compared to controls subjects [56, 57].

PD-related lipid changes also appear to affect lipid raft domains. Thus, similar to AD, lipid rafts from cortical areas (frontal cortex) of PD or iPD subjects revealed an important increase in saturated fatty acids as well as dramatic reductions in DHA and arachidonic acid (AA) [57]. Interestingly, the lipid hallmarks of lipid raft structure, namely sphingolipids and cholesterol, remained unaltered in PD and iPD, suggesting that glycerophospholipids, but not other raft structural lipid constituents, were affected since the early stages of the disease. It is remarkable the observation that these changes in the lipid rafts from frontal cortex of PD and iPD brains were comparable to the alterations observed in AD brains, evoking the possibility that these diseases may have a common neuropathological origin in cortical brain areas.

Lipid abnormalities have also been correlated with protein activities in these microdomains. The relationship between α-Syn and lipid rafts appears to be the consequence of physical interactions with membrane components, including binding to ganglioside GM1 and acidic glycerophospholipids, mainly phosphatidylserine [9, 58, 59]. It has also been established that association of α-Syn with these membrane lipids aids to stabilize an α-helical secondary structure in α-Syn, and that lipid binding is accompanied by an increase in α-helicity which would favour its aggregation [59, 60]. Additional evidences have demonstrated that α-Syn oligomerization and aggregation are promoted by LCPUFA through the activation of retinoic acid X receptor (RXR) and peroxisome proliferator-activated receptor γ2 (PPARγs) [61]. Abnormal α-Syn activities have been shown in mouse models of PD toxicity which display peroxisomal biogenenesis disorders that concur with alterations of LCPUFA levels [61]. Perturbation of protein/lipid integration in lipid rafts, following LCPUFA depletion, leads to α-syn multimerization to form cytotoxic oligomers as a prelude of α-synucleopathies [62, 63]. It is remarkable the observation that lipid changes in raft microstructures are comparable to the alterations in AD [46, 57]. Moreover, similar to Aβ, α-syn regulates LCPUFA, and its expression affects fatty acid composition and brain neutral lipid metabolism in neurons [64, 65].

Signaling Clusters of Lipid Rafts in Neurodegenerative Processes

One of the main causes of neuronal degeneration involves changes in signaling transduction pathways that modify cellular responses. Numerous ionotropic and metabotropic receptors and kinases are involved in neuronal preservation and fuctioning, and neuronal impairment has been attributed in many cases to failure in signaling responses. Due to their particular structure and properties, lipid rafts are a preferential site for signaling molecules crucial for neuronal growth, differentiation, neuroprotection and cellular communication [1, 3]. Thus, signaling molecules cluster in macromolecular complexes (named signalosomes) where transmembrane proteins, lipid-modified proteins or those associated with raft-scaffolding molecules are found, promoting regulation and integration of different intracellular pathways [66].

Lipid matrix of lipid raft microstructures may play an important role in signalosome dynamic and functioning in neurons. In fact, many lipid raft-integrated proteins show lipid-modified structures to provide molecular stabilization. Among others, lipid modifications such as palmitoylation and myristoylation are found in steroid receptors [67, 68], GPI-anchored proteins [69], and kinases [70], as well as raft-scaffolding proteins such as flotillin [71], which are essential for neuronal membrane signaling processes [72]. The importance of lipid/signaling protein associations in lipid rafts indicate that signalosomes may be dynamic microstructures modulated by extracellular or intracellular stimuli [73]. Consequently, alterations in the lipid composition in these microenvironments as a result of neuropathology, such as disruption of homeostatic levels of cholesterol, sphingolipids and LCPUFA, may induce signalosome functional impairments.

In addition to lipid molecules, an important factor for signalosome stability is attributed to lipid raft scaffolding proteins, caveolin 1 and flotillin 1. These proteins play a pivotal role in lipid raft stabilization of signaling proteins, contributing to facilitate their integration and compartimentalization [74] in order to modulate receptor signalling processes [75] relevant for AD, PD and other brain disorders. In this sense, caveolin-1 is actively involved in Aβ formation through, both, its physical association with APP, and modulation of spatial distribution of γ-secretase [76]. Indeed, we have previously demonstrated that estrogen receptors may anchor to caveolin-1 to initiate neuroprotective responses against AD [77]. Also, caveolin-1 plays a functional role in regulating α-Syn actions on neuronal death, a phenomenon which is involved in PD and other synucleinopathies [78]. Furthermore, flotillin-1 interacts with N-methyl-D-aspartate receptors (NMDARs) that mediate excitatory synaptic transmission, and enhances glutamatergic synapse formation, suggesting a role of this protein complex in neurodevelopmental disorders [79, 80].

An interesting concept is that lipid raft structures, through lipid/protein anchoring targets, may represent a "membrane meeting point" where different signaling players converge to initiate neuronal physiological activities, such as neurotrophic, synaptic and neuroprotective activities through still poorly characterized mechanisms of action. In this order of ideas, some data have demonstrated the involvement of clusters of signaling proteins interacting together in lipid rafts as modulators of different neuropathologies. As mentioned before, these microstructures contain APP and APP-derived proteolytic fragments, and participate in recruiting the protein machinery involved in amyloidogenic processes. In this sense, Aβ oligomers are integrated into lipid rafts by means of tyrosine kinase fyn-dependent mechanisms of action. This lipid raft-integrated protein facilitates local actin cytoskeleton rearrengement, thereby inducing Aβ redistribution into the neuronal membrane [81]. Fyn is also an important regulator of APP trafficking in lipid rafts, altering its association with the adaptor protein Dab1 that interacts with amyloid precursor protein (APP), thus modifying Aβ pathological processing [82]. Another kinase acting in lipid rafts is AMP-activated-protein-kinase (AMPK) that inhibits Aβ generation by reducing APP distribution in lipid rafts [83]. Another recent potential target of Aβ aggregates binding is prion protein, PrP(C), which binds in lipid raft structures to amyloid oligomers in a conformation-dependent manner, and reduces Aβ-induced neurotoxicity [2]. Overall, these data reflect a very complex pathway of APP processing and Aβ accumulation which takes place in lipid rafts through still unclear mechanisms.

Lipid raft microenvironments are also suitable places for location of signaling receptors involved in neuronal preservation. Thus, our evidences have demonstrated that estrogen receptor alpha (ERα) and insulin-growth factor-1 receptor (IGF-1R) interact in neuronal lipid rafts promoted by caveolin-1, a phenomenon related to neuroplastic and neuroprotective effects of estradiol against Aβ[84]. An additional component of this complex is a voltage-dependent anion channel (VDAC) which is involved in Aβ-induced toxicity [85], and modulated by estradiol [86]. ERα also interact with metabotropic glutamate receptors (mGluRs) through caveolin-1 in neurons and glia of different brain areas, forming a complex that modulates intracellular signaling involved in neurotransmitter release and alteration of neuronal morphology [87].

Related to PD and α-synucleopathies, as previously mentioned, α-Syn is a main hallmark of PD located in lipid rafts. This protein associates with, both, raft-integrated lipids and proteins, evoking the importance of these microenvironments in α-Syn aggregation and pathological activities. In agreement with this, several lines of evidence have demonstrated that caveolin-1 mediates α-Syn neurotoxicity [88, 89]. Although still awaiting further elucidation, α-Syn is involved in synaptic vesicular recycling [9, 90], and synaptic localization of this protein is mediated by lipid rafts [91]. Moreover, septin 4, a presynaptic lipid raft scaffold protein, has also been shown to interact with α-Syn, acting as a supressor of its neurotoxicity [92].

Overall, these data demonstrate the importance of lipid raft as crucial sites for initiation of both neurotoxic and neuroprotective mechanisms related to AD and PD, acting as strategic membrane microdomains where different molecular players of signal transduction and protein aggregation occurs. Progression of signaling platform activities occurring in these microenvironments may identify additional factors involved in neuronal preservation and functionality. In line with this, proteome studies from cortical lipid rafts in a mouse model of AD have revealed significant alterations in synaptic multiprotein complexes involved in cell signal transduction as a consequence of the pathology, a phenomenon which requires further characterization [93].

Disarrangement of Lipid Raft Constituents in Early Events of AD and PD

Several studies in human AD subjects have pointed out the presence of altered lipid classes in early stages of the disease, though the results are not always consistent. For instance, by using a shotgun lipidomics, Han and colleagues have analyzed over 800 lipid species in control and AD brains and demonstrated decreased levels of sphingomyelin and increased ceramide levels in AD brains [94]. Further studies only revealed a modest reduction in severe AD postmortem brains, and no significant differences in earlier stages of the disease [95]. Importantly, these authors described an extraordinary depletion of sulfatide levels in all analyzed brain regions of AD subjects (up to 93% in grey mater) and these alterations were already observed in the earliest stages of the disease [94].

On the other hand, a differential reduction of gangliosides in the grey matter of early onset AD subjects have been observed compared not only to control individuals but also to late onset AD [96]. A more recent study found elevated levels of GM1 and GM2 in lipid raft fractions of the temporal and frontal cortex of AD brains [33]. Reductions in LCPUFA have also been demonstrated to occur not only in advanced AD patients but also in early stages of the disease [97]. Moreover, recent evidences from our group have revealed that many of these lipid alterations in brain tissues have also a reflection on neuronal lipid rafts, and represent very early events in Alzheimer's disease (unpublished results). In the case of AD, these effects are preferentially observed in frontal and entorhinal cortices, two brain areas which are particularly affected during the progression of AD neuropathology. Strikingly, we have observed severe alterations in the lipid matrix of lipid rafts from frontal cortex of Parkinson's disease brains, which were already present in incidental PD [57]. In line with this concept, early stages of PD pathology have also been associated with MCI in the first steps of the progression of the disease [98, 99]. Importantly, it should be noted that peroxidation products of DHA and AA, which are very susceptible to lipid peroxidation due to their high contents in double bounds, are elevated in AD and PD brains, not only in advanced patients but also in early stages of the diseases [99, 100, 101]. This issue raises the hypothesis that DHA and AA loss can be ascribed to the increased oxidative stress, which is one of the accepted causes for AD and PD [56, 99, 101].

The pathophysiological consequences of lipid alterations in lipid rafts during first stages of AD and PD may have a mechanistic connection with physiological disarrangements of signaling platforms in these microstructures. In agreement with this hypothesis, our results in lipid rafts of cortical areas of early AD at stages I/II [102] have demonstrated that the modifications in the lipid matrix of lipid rafts have important consequences in, both, physical properties and amyloidogenic processing of APP. Noticeably, these impairments were of similar magnitude than those observed in severe AD (unpublished results), observing significant alterations in APP, BACE and γ-secretase associations that affect β-amyloidogenesis. To our knowledge, these evidences represent the first demonstration that alterations in brain lipid homeostasis in early stages of AD induce molecular changes that may affect the pathogenesis of this disease. Taking into account that lipid alterations in lipid rafts are a common starting parameter of AD and PD [46, 57], it is conceivable that similar molecular disorders may occur in early events of these two pathologies, thereby contributing to neuropathological development. Further studies may confirm this hypothesis.

Nonetheless, it should be stated that early events of lipid rafts alterations associated with AD or PD neurodegenerations, must be clearly distinguished from those occurring during normal aging. Indeed, our recent data indicates that aging is accompanied by subtle changes in the lipid composition of lipid rafts, a process that we have coined as "lipid raft aging" [51]. Some of these changes occur in lipid species (such as sulfatides or saturated fatty acids) that are severely affected during AD and PD neurodegenerations. This naturally-occurring modification is considerably accelerated and agravated by the presence of AD fenotype, at least in APP/PS1 transgenic mice.

The Challenge of Novel Biomarkers Derived from Lipid Raft Alterations as a Consequence of Neuropathologies

AD and PD are incurable diseases that have a high prevalence in the population in developed countries [103]. Although important efforts have been made in order to obtain more accurate and specific treatements, conclusive diagnostic parameteres are evidenced in advanced stages of the pathology, with a high degree of brain damage, thereby reducing treatment's effectiveness. Consequently, a main requirement is the search of novel biomarkers detectable by non-invasive methods to identify at-risk individuals during presymptomatic stages of AD, thereby obtaining insights allowing the development of more effective therapies.

The diagnostic guidelines of AD presently accepted include standard diagnostic protocols, consisting of neurological tests, neuroimaging, clinical, genetic and neurochemical criteria [104, 105]. Based upon these diagnostic parameters, three different stages of AD have been identified: preclinical AD, mild cognitive impairment (MCI) and AD with dementia. However, these biomarkers do not fulfill the diagnostic criteria of sensitivity and specificity in early stages of AD. A recent updating of diagnosis have added cerebrospinal fluid (CSF) biomarkers, with the aim of obtaining a more clear scenario ongoing molecular alterations in AD patients. Validated CSF diagnostic biomarkers for severe AD include the determination of amyloid beta 1-42, total tau (T-tau) and phospho-tau (P-Tau) levels, thereby reflecting the main hallmarks of AD, senile plaque formation by Aβ aggregation, and neurofibrillary tangle formation by Tau hyperphosphorylation. Some cross-sectional and longitudinal studies have analyzed the relation between these CSF AD biomarkers and preclinical AD. The studies concluded that there are significant changes in the levels of these biomarkers between healthy elderly and AD patients, although the results are inconclusive on patients with cognitive normal at-risk individuals [106]. In this sense, standardization of analytical procedures should be set up, as a prerequisite for trustworthy analysis. In particular, related to MCI, which is associated with subtle cognitive deficits with an increase risk of developing dementia, diagnostic criteria lack consensual support [107]. Nevertheless, improvement of biomarkers in MCI require prospective long-term studies to predict AD development in individuals within a decade. Indeed, the risk to develop AD increases three-fold in patients with MCI as compared to healthy controls [108], and a main drawback is that MCI is a very heterogeneous disorder, leading to the progression of a variety of dementia [109].

Biomarkers for PD diagnosis remains challenging since, to date, no conclusive results have been obtained for early stages of this disease [110]. Genetic, imaging and even olfactory tests are analyzed in the aim of tackling early diagnostic aspects of PD [111, 112]. Non-motor symptoms of parkinsonism precede severe PD, as observed by autopsy observations where neuronal impairment is observed in non-nigrostriatal neurons [113]. Moreover, the pathology of PD has also been associated with MCI as an early manifestation of the disease [98]. In agreement with this, recent measures of Aβ1-42, T-Tau and P-Tau, together with α-Syn, in CSF of PD patients revealed a prognostic and diagnostic potential in incidental PD, supporting the idea that these CSF biomarkers may be useful also for prediction of PD diagnosis [114]. Other promising CSF markers are under study related to oxidative stress [115] and lipid alterations [116]. However, taking into account the multifactorial aspects of PD with complex diagnostic strategies in early stages that share etiological factors with other neurodegenerative diseases, it is plausible that a future accurate approach may be include a plethora of CSF biomarkers in multivariate analysis in order to reach sensitivity and specificity.

Thus, a major challenge is the development of biomarkers that may be utilized for prognosis and disease progression, in the aim of identifying specific cognitively normal at-risk individuals for AD and other dementia. In this order of ideas, we propose that detection of lipid raft changes underlying neuronal impairment could lead to the development of early specific diagnostic approaches. Some encouraging results are in progress related to alterations in lipid homeostasis and signal transduction in neurons [117]. We proposed that lipid disarrangements as a consequence of neuronal damage may be reflected in peripheral fluids. Amongst them, indicators of lipoperoxidation and levels of metabolites resulting from oxidative damage of lipids in these microstructures, as well as activities of enzymes involved in degradation of constitutive lipids from lipid rafts (such SMases and PLA2 phospholipases), may be assessed in CSF samples of MCI subjects.

Furthermore, several protein candidates of lipid raft derangements in early stages of AD may include CSF detection of VDAC, a raft-integrated porine involved in Aβ toxicity mechanisms [85] which has been found associated with γ-secretase [118]. Lipid raft scaffolding proteins may also be worth to be analyzed in these fluids, as a potential mirror of neuronal anomalies occurring in cortical neurons. Toxic post-translational modification of raft-integrated proteins as a consequence of aberrant assemblies/disassemblies levels may also serve to evaluate the degree of neuronal lesion [119]. In line with this, it is plausible that lipid raft anomalies may enhance NO production, and protein S-nitrosylation (covalent modification of thiol groups), and carbonylation, thereby contributing to protein missfolding and neurodegeneration [120, 121]. Thus, some data have revealed that BACE1 is highly modified by NO production, and its expression is exposed to dual regulation depending on either S-nitrosative or oxidative stress, and the presence of 4-hydroxynonenal [122, 123]. This reveals a molecular link between oxidative toxicity and APP processing to generate Aβ [124].

Overall, we proposed that a plausible scenario of future diagnostic tools at pre-clinical stages of AD and PD may consist in a multivariate analysis of selected lipid, protein and biometal parameters associated with neuronal lipid raft impairments, in cerebrospinal and peripheral fluids. These approaches may likely provide a disease-specific neuropathological signature at different stages of these desvasting neurodegenerative diseases.

Conclusion

Prevalence of AD and PD neuropathologies is increasing in parallel with an increase in longevity, in particular in western countries. Since these diseases are incurable, this phenomenon poses a serious socio-economic trouble, and an urgent public health challenge. Present treatment's accuracy is subject to pertinent diagnostic parameters which are frequently evidenced in advanced stages of brain detriment, thereby reducing efectiveness. Thus, part of the strategy for more efficient therapeutic strategies requires the development of novel diagnostic biomarkers to predict, in at-risk individuals, future development of AD or PD. In this sense, lipid rafts from brain cortex undergo age-dependent alterations (coined as "lipid raft ageing") which are an indication of the normal biological decay [51]. This process, characterized by severe loss of lipid raft homeostasis, protein disarrangements, and functional impairment, is exacerbated in AD and PD. Apparently, the evolution from normal to pathological lipid raft ageing occurs gradually, and the transition may presage the initiation of the neuropathology. Therefore, detection of these early events of the disease at the cerebrospinal fluid level, as a mirror of the neuronal scenario, may provide novel biomarkers for early stages of AD and PD neurodegeneration. CSF measurement of selected lipid and protein species require multivariate approaches to assess with accuracy the degree of correlation and discrimination of these biomarkers between different neuropatologies in pre-symptomatic subjects. Future investigations will pave the way for optimal diagnostic tools to identify the first presymptomatic signs of AD and PD, in the aim of improving beneficial treatment results.

Acknowledgments

This work was supported by grants SAF2010-22114-C02-01 and SAF2010-22114-C02-02 (Spanish Ministry of Economy and Competitiveness).

References

[1] Allen, J. A., Halverson-Tamboli, R. A. & Rasenick, M. M. (2007). Lipid raft microdomains and neurotransmitter signalling. *Nat Rev Neurosci.*, *8*(2), 128-40

[2] Rushworth, J. V. & Hooper, N. M. (2010). Lipid rafts: Linking Alzheimer's amyloid-b production, aggregation and toxicity at neuronal membranes. *Int J Alzheimer s Dis.*, *2011*, 603052-603064

[3] Tsui-Pierchala, B. A., Encinas, M., Milbrandt, J. & Johnson, E. MJr. (2002). Lipid rafts in neuronal signaling and function. *Trends Neurosci.*, *25*, 412-417

[4] Guirland, C. & Zheng, J. Q. (2007). Membrane lipid rafts and their role in axon guidance. *Adv Exp Med Biol.*, *621*, 144-55

[5] Michel, V. & Bakovic, M. (2007). Lipid rafts in health and disease. *Biol. Cell*, *99*, 129-140

[6] Schengrund, C. L. (2010). Lipid rafts: keys to neurodegeneration. *Brain Res Bull.*, *82*(1-2), 7-17

[7] Ma, D. W. (2007). Lipid mediators in membrane rafts are important determinants of human health and disease. *Appl Physiol Nutr Metab.*, *32*(3), 341-50

[8] Gellermann, G. P., Appel, T. R., Tannert, A., Radestock, A., Hortschansky, P., Schroeckh, V., Leisner, C., Lütkepohl, T., Shtrasburg, S., Röcken, C., Pras, M., Linke, R. P., Diekmann, S. & Fändrich, M. (2005). Raft lipids as common components of human extracellular amyloid fibrils. *Proc Natl Acad Sci*, *102*, 6297-6302

[9] Kubo, S., Nemani, V. M., Chalkley, R. J., Anthony, M. D., Hattori, N., Mizuno, Y., Edwards, R. H. & Fortin, D. L. (2005). A combinatorial code for the interaction of alpha-synuclein with membranes. *J Biol Chem.*, *280*(36), 31664-72

[10] Marin, R. (2011). Signalosomes in the brain: relevance in the development of certain neuropathologies such as Alzheimer's disease. *Front Physiol.*, *2*(23), 1-3

[11] Hicks, D. A., Nalivaeva, N. N. & Turner, A. (2012). Lipid rafts and Alzheimer's disease: protein-lipid interactions and perturbation of signaling. *Front. Physiol.*, *3*(189), 1-18

[12] Duyckaerts, C. & Dickson, D. W. (2003). Neuropathology of Alzheimer's disease, D.W. Dickson (Ed.), *Neurodegeneration: The Molecular Pathology of Dementia and Movement Disorders*, ISN Neuropath Press, Basel 47–65

[13] Lowe, J., Mirra, S. S., Hyman, B. & Dickson, D. W. (2008). Ageing and Dementia, S. Love, D.N. Louis, D.W. Ellison (Eds.), *Greenfield's Neuropathology*, Hodder Arnold, London., *1*, 1031-1152

[14] Abad Rodriguez, J., Ledesma, M. D., Craessaerts, K., Perga, S., Medina, M., Delacourte, A., Dingwall, C., De Strooper, B. & Dotti, C. G. (2004). Neuronal membrane cholesterol loss enhances amyloid peptide generation. *J Cell Biol.*, *167*, 953-960

[15] Allinson, T. M. J., Parkin, E. T., Turner, A. J. & Hooper, N. M. (2003). ADAMs family members as amyloid precursor protein α-secretases. *Journal of Neuroscience Research*, vol. *74*, no. 3, 342–352

[16] Schneider, A., Rajendran, L., Honsho, M., Gralle, M., Donnert, G., Wouters, F., Hell, S. W. & Simons, M. (2008). Flotillin-dependent clustering of the amyloid precursor protein regulates its endocytosis and amyloidogenic processing in neurons. *J Neurosci.*, *28*(11), 2874-82

[17] Chen, T. Y., Liu, P. H., Ruan, C. T., Chiu, L. & Kung, F. L. (2006). The intracellular domain of amyloid precursor protein interacts with flotillin-1, a lipid raft protein. *Biochem Biophys Res Commun.*, *342*(1), 266-72

[18] Ehehalt, R., Keller, P., Haass, C., Thiele, C. & Simons, K. (2003). Amyloidogenic processing of the Alzheimer beta-amyloid precursor protein depends on lipid rafts. *J. Cell Biol.*, *160*, 113-123
[19] Kalvodova, L., Kahya, N., Schwille, P., Ehehalt, R., Verkade, P., Drechsel, D. & Simons, K. (2005). Lipids as modulators of proteolytic activity of BACE, involvement of cholesterol, glycosphingolipids & anionic phospholipids in vitro. *J Biol Chem*, *280*, 36815-36823
[20] Vetrivel, K. S. & Thinakaran, G. (2010). Membrane rafts in Alzheimer's disease beta-amyloid production. *Biochim Biophys Acta.*, *1801*(8), 860-867
[21] Kapoor, A., Hsud, W. M., Wang, B. J., Wu, G. H., Lin, T. Y., Lee, S., Yen, Ch. T. & Liayo Y. F. (2010). Caveolin1 regulates secretase mediated AβPP processing by modulating spatial distribution of secretase in membrane. *J Alzheimer's Dis.*, *22*, 423-442
[22] Harris, B., Pereira, I. & Parkin, E. (2009). Targeting ADAM10 to lipid rafts in neuroblastoma SH-SY5Y cells impairs amyloidogenic processing of the amyloid precursor protein. *Brain Res.*, *1296*, 203-15
[23] Sakurai, T., Kaneko, K., Okuno, M., Wada, K., Kashiyama, T., Shimizu, H., Akagi, T., Hashikawa, T. & Nukina, N. (2008). Membrane microdomain switching: a regulatory mechanism of amyloidprecursor protein processing. *J Cell Biol.*, *183*(2), 339-52
[24] Askarova, S., Yang, X. & Lee, J. C. (2011). Impacts of membrane biophysics in Alzheimer's disease: from amyloid precursor protein processing to Aβ Peptide-induced membrane changes. *Int J Alzheimers Dis. doi*, 10.4061/2011/134971.
[25] Diaz, M., Fabelo, N. & Marin, R. (2012). Genotype-induced changes in biophysical properties of frontal cortex lipid raft from APP/PS1 transgenic mice. *Front. Physiol.*, *3*(3), 1-10
[26] Williams, T. L. & Serpell, L. C. (2011). Membrane and surface interactions of Alzheimer's Aβ peptide-insights into the mechanism of cytotoxicity. *FEBS J.*, *278*(20), 3905-17
[27] Kakio, A., Nishimoto, S., Kozutsumi, Y. & Matsuzaki, K. (2003). Formation of a membrane-active form of amyloid beta-protein in raft-like model membranes. *Biochem Biophys Res Commun.*, *303*(2), 514-8
[28] Hayashi, H., Kimura, N., Yamaguchi, H., Hasegawa, K., Yokoseki, T., Shibata, M., Yamamoto, N., Michikawa, M., Yoshikawa, Y., Terao, K., Matsuzaki, K., Lemere, C. A., Selkoe, D. J., Naiki, H. & Yanagisawa, K. (2004). A seed for Alzheimer amyloid in the brain. *J Neurosci.*, *24*(20), 4894-90

[29] Bucciantini, M., Nosi, D., Forzan, M., Russo, E., Calamai, M., Pieri, L., Formigli, L., Quercioli, F., Soria, S., Pavone, F., Savistchenko, J., Melki, R. & Stefani, M. (2012). Toxic effects of amyloid fibrils on cell membranes: the importance of ganglioside GM1. *FASEB J.*, *26*(2), 818-31

[30] Peters, I., Igbavboa, U., Schütt, T., Haidari, S., Hartig, U., Rosello, X., Böttner, S., Copanaki, E., Deller, T., Kögel, D., Wood, W. G., Müller, W. E. & Eckert, G. P. (2009). The interaction of beta-amyloid protein with cellular membranes stimulates its own production. *Biochim Biophys Acta*, *1788*(5), 964-72

[31] Malchiodi-Albedi, F., Contrusciere, V., Raggi, C., Fecchi, K., Rainaldi, G., Paradisi, S., Matteucci, A., Santini, M. T., Sargiacomo, M., Frank, C., Gaudiano, M. C. & Diociaiuti, M. (2010). Lipid raft disruption protects mature neurons against amyloid oligomer toxicity. *Biochimica Biophysica Acta*, *1802*(4), 406–415

[32] Grimm, M. O., Rothhaar, T. L. & Hartmann, T. (2012). The role of APP proteolytic processing in lipid metabolism. *Exp Brain Res.*, *217*(3-4), 365-75

[33] Molander-Melin, M., Blennow, K., Bogdanovic, N., Dellheden, B., Månsson, J. E. & Fredman, P. (2005). Structural membrane alterations in Alzheimer brains found to be associated with regional disease development; increased density of gangliosides GM1 and GM2 and loss of cholesterol in detergent-resistant membrane domains. *J Neurochem.*, *92*(1), 171-82

[34] Ariga, T., McDonald, M. P. & Yu, R. K. (2008). Role of ganglioside metabolism in the pathogenesis of Alzheimer's disease-a review. *J Lipid Res.*, *49*(6), 1157-75

[35] Chan, R. B., Oliveira, T. G., Cortes, E. P., Honig, L. S., Duff, K. E., Small, S. A., Wenk, M. R., Shui, G. & Di Paolo, G. (2012). Comparative lipidomic analysis of mouse and human brain with Alzheimer disease. *J Biol Chem.*, *287*(4), 2678-88

[36] Soreghan, B., Thomas, S. N. & Yang, A. J. (2003). Aberrant sphingomyelin/ ceramide metabolic-induced neuronal endosomal/ lysosomal dysfunction: potential pathological consequences in age-related neurodegeneration. *Adv Drug Deliv Rev.*, *55*, 1515-1524

[37] Tong, M. & de la Monte, S. M. (2009). Mechanisms of ceramide-mediated neurodegeneration. *J Alzheimers Dis.*, *16*(4), 705-14

[38] Jana, A., Hogan, E. L. & Pahan, K. (2009). Ceramide and neurodegeneration: susceptibility of neurons and oligodendrocytes to cell damage and death. *J Neurol Sci.*, *278*(1-2), 5-15

[39] Yang, X., Askarova, S. & Lee, J. C. (2010). Membrane biophysics and mechanics in Alzheimer's disease. *Mol Neurobiol.*, *41*(2-3), 138-48

[40] Wood, W. G., Schroeder, F., Igbavboa, U., Avdulov, N. A. & Chochina, S. V. (2002). Brain membrane cholesterol domains, aging and amyloid beta-peptides. *Neurobiol. Aging.*, *23*(5), 685-694

[41] Grösgen, S., Grimm, M. O., Friess, P. & Hartmann, T. (2010). Role of amyloid beta in lipid homeostasis. *Biochim Biophys Acta.*, *1801*(8), 966-74

[42] Simons, M., Keller, P., Dichgans, J. & Schulz, J. B. (2001). Cholesterol and Alzheimer's disease: is there a link? *Neurology*, *57*, 1089-1093

[43] Kojro, E., Gimpl, G., Lammich, S., Marz, W. & Fahrenholz, F. (2001). Low cholesterol stimulates the nonamyloidogenic pathway by its effect on the α-secretase ADAM 10. *Proc Natl Acad Sci USA.*, *98*, 5815-5820

[44] Cecchi, C., Nichino, D., Zampagni, M., Bernacchioni, C., Evangelisti, E., Pensalfini, A., Liguri, G., Gliozzi, A., Stefani, M. & Relini, A. (2009). A protective role for lipid raft cholesterol against amyloid-induced membrane damage in human neuroblastoma cells. *Biochim Biophys Acta*, *1788*, 2204-2216

[45] Kolsch, H., Heun, R., Jessen, F., Popp, J., Hentschel, F., Maier, W. & Lutjohann, D. (2010). Alterations of cholesterol precursor levels in Alzheimer's disease. *Biochim. Biophys. Acta*, *1801*, 945–950

[46] Martin, V., Fabelo, N., Santpere, G., Puig, B., Marín, R., Ferrer, I. & Díaz, M. (2010). Lipid alterations in lipid rafts from Alzheimer's disease human brain cortex. *J Alzheimer's Dis.*, *19*(2), 489-502

[47] Young, G. & Conquer, J. (2005). Omega-3 fatty acids and neurophsychiatric disorders. *Reprod Nutr Dev*, *45*(1), 1-28

[48] Wassall, S. R. & Stillwell, W. (2009). Polyunsaturated fatty acid cholesterol interactions: domain formation in membranes. *Biochim Biophys Acta.*, *1788*(1), 24-32

[49] Soderberg, M., Edlund, C., Kristensson, K. & Dallner, G. (1991). Fatty acid composition of brain phospholipids in aging and in Alzheimer's disease. *Lipids*, *26*, 421-425

[50] Plourde, M. & Cunnane, S. C. (2007). Extremely limited synthesis of long chain polyunsaturates in adults: implications for their dietary essentiality and use as supplements. *Appl Physiol Nutr Metab.*, *32*, 619-634

[51] Fabelo, N., Martín, V., Marin, R., Santpere, G., Aso, E., Ferrer, I. & Diaz, M. (2012). Evidence for premature lipid raft aging in APP/PS1 double-transgenic mice, a model of familial Alzheimer disease. *J Neuropathol Exp Neurol.*, *71*(10), 868-81

[52] Goedert, M. (2001). Alpha-synuclein and neurodegenerative diseases. (2001) *Nat Rev Neurosci.*, *2*(7), 492-501

[53] Iwatsubo, T. (2003). Aggregation of alpha-synuclein in the pathogenesis of Parkinson's disease. *J Neurol.*, *250* (3), 1-4

[54] Julien, C., Berthiaume, L., Hadj-Tahar, A., Rajput, A. H., Bédard, P. J., Di Paolo, T., Julien, P. & Calon, F. (2006). Postmortem brain fatty acid profile of levodopa-treated Parkinson disease patients and parkinsonian monkeys. *Neurochem Int.*, *48*(5), 404-14

[55] Bousquet, M., Calon, F. & Cicchetti, F. (2011). Impact of ω-3 fatty acids in Parkinson's disease. *Ageing Res Rev.*, *10*(4), 453-63

[56] Dalfó, E., Portero-Otín, M., Ayala, V., Martínez, A., Pamplona, R. & Ferrer, I. (2005). Evidence of oxidative stress in the neocortex in incidental Lewy body disease. *J Neuropathol Exp Neurol.*, *64*(9), 816-30

[57] Fabelo, N., Martín, V., Santpere, G., Marin, R., Torrent, L., Ferrer, I. & Díaz, M. (2011). Severe alterations in lipid composition of frontal cortex lipid rafts from Parkinson's disease and incidental Parkinson's disease. *Mol Med.*, *17*(9-10), 1107-18

[58] Davidson, W. S., Jonas, A., Clayton, D. F. & George, J. M. (1998). Stabilization of alpha-synuclein secondary structure upon binding to synthetic membranes. *J Biol Chem.*, *273*(16), 9443-50

[59] Martinez, Z., Zhu, M., Han, S. & Fink, A. L. (2007). GM1 specifically interacts with alpha-synuclein and inhibits fibrillation. *Biochemistry*, *46*(7), 1868-77

[60] Cole, N. B., Murphy, D. D., Grider, T., Rueter, S., Brasaemle, D. & Nussbaum, R. L. (2002). Lipid droplet binding and oligomerization properties of the Parkinson's disease protein alpha-synuclein. *J Biol Chem.*, *277*(8), 6344-52

[61] Yakunin, E., Moser, A., Loeb, V., Saada, A., Faust, P., Crane, D. I., Baes, M. & Sharon, R. (2010). Alpha-Synuclein abnormalities in mouse models of peroxisome biogenesis disorders. *J Neurosci Res.*, *88*(4), 866-76

[62] Perrin, R. J., Woods, W. S., Clayton, D. F. & George, J. M. (2001). Exposure to long chain polyunsaturated fatty acids triggers rapid multimerization of synucleins. *J Biol Chem.*, *276*(45), 41958-62

[63] Assayag, K., Yakunin, E., Loeb, V., Selkoe, D. J. & Sharon, R. (2007). Polyunsaturated fatty acids induce alpha-synuclein-related pathogenic changes in neuronal cells. *Am J Pathol.*, *171*(6), 2000-11

[64] Sharon, R., Bar-Joseph, I., Frosch, M. P., Walsh, D. M., Hamilton, J. A. & Selkoe, D. J. (2003). Altered fatty acid composition of dopaminergic neurons expressing a-synuclein and human brains with a-synucleinopathies. *Neuron*, *37*, 583-595

[65] Barceló-Coblijn, G., Golovko, M. Y., Weinhofer, I., Berger, J. & Murphy, E. J. (2007). Brain neutral lipids mass is increased in alpha-synuclein gene-ablated mice. *J Neurochem.*, *101*(1), 132-41

[66] Lisanti, M. P., Scherer, P. E., Tang, Z. & Sargiacomo, M. (1994). Caveolae, caveolin and caveolin-rich membrane domains: a signalling hypothesis. *Trends Cell Biol.*, *4*, 231-235

[67] Marino, M., Ascenzi, P. & Acconcia, F. (2006). S-palmitoylation modulates estrogen receptor alpha localization and functions. *Steroids*, *71*, 298-303

[68] Pedram, A., Razandi, M., Sainson, R. C., Kim, J. K., Hughes, C. C. & Levin, E. R. (2007). A conserved mechanism for steroid receptor translocation to the plasma membrane. *J Biol Chem.*, *282*, 22278-22288

[69] Murakami, Y., Siripanyapinyo, U., Hong, Y., Kang, J. Y., Ishihara, S., Nakakuma, H., Yusuke, M. & Kinoshita, T. (2003). PIG-W Is Critical for Inositol Acylation but Not for Flipping of Glycosylphosphatidylinositol-Anchor. *Mol Biol Cell.*, *14*, 4285-4295

[70] Sato, I., Obata, Y., Kasahara, K., Nakayama, Y., Fukumoto, Y., Yamasaki, T., Yokohama, K. K., Saito, T. & Yamaguchi, N. (2009). Differential trafficking of Src, Lyn, Yes and Fyn is specified by the state of palmitoylation in the SH4 domain, *J Cell Sci.*, *122*(Pt 7), 965-975

[71] Morrow, I. C., Rea, S., Martin, S., Prior, I. A., Prohaska, B., Hancock, J. F., James, D. E. & Parton, R. G. (2002). Flotillin-1/Reggie-2 traffics to surface raft domains via a novel Golgi-independent pathway. *J Biol Chem.*, *277*, 48834-48841

[72] Meitzen, J., Luoma, J. I., Boulware, M. I., Hedges, V. L., Peterso, B. M., Tuomela, K., Britson, K. A. & Mermelstein, P. G. (2013). Palmitoylation of estrogen receptors is essential for neuronal membrane signaling. *Endocrinology*, doi, 10.1210/en.2013-1172

[73] Lingwood, D. & Simons, K. (2010). Lipid rafts as a membrane organizing principle. *Science*, *327*, 46-50

[74] Parton, R. G. (2003). Caveolae-form ultrastructure to molecular mechanisms. *Nat. Rev. Mol. Cell Biol.*, *4*, 162-167

[75] Lajoie, P., Goetz, J. G., Dennis, J. W. & Nabi, I. R. (2009). Lattices, rafts, and scaffolds: domain regulation of receptor signaling at the plasma membrane. *J. Cell Biol.*, *185*, 381-385

[76] Kapoor, A., Hsud, W. M., Wang, B. J., Wu, G. H., Lin, T. Y., Lee, S., Yen, ChT. & Liao, Y. F. (2010). Caveolin1 regulates secretase mediated AbPP processing by modulating spatial distribution of secretase in membrane. *J. Alzheimer's Dis.*, *22*, 423-442

[77] Marin, R., Ramirez, C. M., Morales, A., Gonzalez, M., Alonso, R. & Diaz, M. (2008). Modulation of Ab-induced neurotoxicity by estrogen receptor alpha and other associated proteins in lipid rafts. *Steroids*, *73*, 992-996

[78] Madeira, A., Yang, J., Zhang, X., Vikeved, E., Nilsson, A., Andrén, P. E. & Svenningsson, P. (2011). Caveolin-1 interacts with alpha-synuclein and mediates toxic actions of cellular alpha-synuclein overexpression. *Neurochem Int.*, *59*(2), 280-9

[79] Swanwick, C. C., Shapiro, M. E., Yi, Z., Chang, K. & Wenthold, R. J. (2009). NMDA receptors interact with flotillin-1 and -2, lipid raft-associated proteins. *FEBS Lett.*, *583*(8), 1226-30

[80] Swanwick, C. C., Shapiro, M. E., Vicini, S. & Wenthold, R. J. (2010). Flotillin-1 promotes formation of glutamatergic synapses in hippocampal neurons. *Dev Neurobiol*, *70*(13), 875-83

[81] Williamson, R., Usardi, A., Hanger, D. P. & Anderton, B. H. (2008). Membrane-bound beta-amyloid oligomers are recruited into lipid rafts by a fyn-dependent mechanism. *FASEB J.*, *22*(5), 1552-9

[82] Minami, S. S., Hoe, H. S. & Rebeck, G. W. (2011). Fyn kinase regulates the association between amyloid precursor protein and Dab1 by promoting their localization to detergent-resistant membranes. *J Neurochem.*, *118*(5), 879-90

[83] Won, J. S., Im, Y. B., Kim, J., Singh, A. K. & Singh, I. (2010). Involvement of AMP-activated-protein-kinase (AMPK) in neuronal amyloidogenesis. *Biochem Biophys Res Commun.*, *399*(4), 487-91

[84] Marin, R., Diaz, M., Alonso, R., Sanz, A., Arevalo, M. A. & Garcia-Segura, L. M. (2009). Role of estrogen receptor in membrana-initiated signaling in neural cells: interaction with IGF-1 receptor. J. Steroid Biochem. *Mol. Biol.*, *114*, 2-7

[85] Marin, R., Ramírez, C. M., González, M., González-Muñoz, E., Zorzano, A., Camps, M., Alonso, R. & Díaz, M. (2007). Voltage-dependent anion channel (VDAC) participates in amyloid beta-induced

toxicity and interacts with plasma membrane estrogen receptor a in septal and hippocampal neurons. *Mol. Memb. Biol.*, *24*, 148-60

[86] Herrera, J. L., Diaz, M., Hernández-Fernaud, J. R., Salido, E., Alonso, R., Fernández, C., Morales, A. & Marin, R. (2011). Voltage-dependent anion channel as a resident protein of lipid rafts: post-transductional regulation by estrogens and involvement in neuronal preservation against Alzheimer's disease. *J Neurochem.*, *116*(5), 820-7

[87] Micevych, P. & Christensen, A. (2012). Membrane-initiated estradiol actions mediate structural plasticity and reproduction. *Front Neuroendocrinol.*, *33*(4), 331-41

[88] Hashimoto, M., Takenouchi, T., Rockenstein, E. & Masliah, E. (2003). Alpha-synuclein up-regulates expression of caveolin-1 and down-regulates extracellular signal-regulated kinase activity in B103 neuroblastoma cells: role in the pathogenesis of Parkinson's disease. *J. Neurochem.*, *85*, 14681-1479

[89] Benarroch, E. E. (2007). Lipid rafts, protein scaffolds, and neurologic disease. *Neurology*, *69*, 1635-1639

[90] Sidhu, A., Wersinger, C. & Vernier, P. (2004). Does alpha-synuclein modulate dopaminergic synaptic content and tone at the synapse? *FASEB J.*, *18*, 637-647

[91] Fortin, D. L., Troyer, M. D., Nakamura, K., Kubo, S., Anthony, M. D. & Edwards, R. H. (2004). Lipid rafts mediate the synaptic localization of alpha-synuclein. *J Neurosci.*, *24*, 6715-6723

[92] Ihara, M., Yamasaki, N., Hagiwara, A., Tanigaki, A., Kitano, A., Hikawa, R., Tomimoto, H., Noda, M., Takanashi, M., Mori, H., Hattori, N., Miyakawa, T. & Kinoshita, M. (2007). Setp4, a component of presynaptic scaffold and Lewy bodies, is required for the suppression of alpha-synuclein neurotoxicity. *Neuron*, *53*, 519-533

[93] Chadwick, W., Brenneman, R., Martin, B. & Maudsley, S. (2010). Complex and multidimensional lipid raft alterations in a murine model of Alzheimer's disease. *Int J Alzheimers Dis.*, *2010*, 604-792

[94] Han, X., Rozen, S., Boyle, S. H., Hellegers, C., Cheng, H., Burke, J. R., Welsh-Bohmer, K. A., Doraiswamy, P. M. & Kaddurah-Daouk, R. (2011). Metabolomics in early Alzheimer's disease: identification of altered plasma sphingolipidome using shotgun lipidomics. *PLoS ONE*, *6*, e21643. doi, 10.1371/journal.pone.0021643

[95] Han, X., Holtzman, D. M., McKeel, D. W., Kelley, J. & Morris, J. C. (2002). Substantial sulfatide deficiency and ceramide elevation in very

early Alzheimer's disease: potential role in disease pathogenesis. *Journal of Neurochemistry*, *82*(4), 809-818

[96] Gottfries, C. G, Karlsson, I. & Svennerholm, L. (1996). Membrane components separate early-onset Alzheimer's disease from senile dementia of the Alzheimer type. *Int Psychogeriatr.*, *8*(3), 365-72

[97] Janvin, C. C., Larsen, J. P., Aasrland, D. & Hugdahl, K. (2006). Subtypes of mild cognitive impairment in Parkinson's disease: progression to dementia. *Mov Disord.*, *21*(9), 1343-9

[98] Poletti, M., Emre, M. & Bonuccelli, U. (2011). Mild cognitive impairment and cognitive reserve in Parkinson's disease. *Parkinsonism Relat Disord.*, *17*(8), 579-86

[99] Markesbery, W. R., Kryscio, R. J., Lovell, M. A. & Morrow, J. D. (2005). Lipid peroxidation is an early event in the brain in amnestic mild cognitive impairment. *Ann Neurol.*, *58*(5), 730-735

[100] Montine, T. J. & Morrow, J. D. (2005). Fatty acid oxidation in the pathogenesis of Alzheimer's disease. *American Journal of Pathology*, *166*(5), 1283-1289

[101] Tsang, A. H. & Chung, K. K. (2009). Oxidative and nitrosative stress in Parkinson's disease. *Biochim Biophys Acta.*, *1792*(7), 643-50

[102] Braak, H. & Braak, E. (1991). Neuropathological stageing of Alzheimer-related changes. *Acta Neuropathol*, *82*, 239-259

[103] Dorsey, E. R., Constantinescu, R., Thompson, J. P., Biglan, K. M., Holloway, R. G., Kieburtz, K., Marshall, F. J., Ravina, B. M., Schifitto, G., Siderowf, A. & Tanner, C. M. (2007). Projected number of people with Parkinson disease in the most populous nations, 2005 through 2030. *Neurology*, *68*, 384–386

[104] Dubois, B., Feldman, H. H., Jacova, C., Dekosky, S. T., Barberger-Gateau, P., Cummings, J., et al. (2007). Research criteria for the diagnosis of Alzheimer's disease: revising the NINCDS–ADRDA criteria. *Lancet Neurol*, *6*(8), 734–746

[105] Forlenza, O. V., Diniz, B. S. & Gattaz, W. F. (2010). Diagnosis and biomarkers of predementia in Alzheimer's disease. *BMC Med*, *8*, 89-94

[106] Rosén, C., Hansson, O., Blennow, K. & Zetterberg, H. (2013). Fluid biomarkers in Alzheimer's disease - current concepts. *Mol Neurodegener.*, *8*, 20-31

[107] Busse, A., Bischkopf, J., Riedel-Heller, S. G. & Angermeyer, M. C. (2003). Mild cognitive impairment: prevalence and predictive validity according to current approaches. *Acta Neurol Scand*, *108*(2), 71-81

[108] Bennett, D. A., Wilson, R. S., Schneider, J. A., Evans, D. A., Beckett, L. A., Aggarwal, N. T., Barnes, L. L., Fox, J. H. & Bach, J. (2002). Natural history of mild cognitive impairment in older persons. *Neurology*, *59*, 198-205

[109] Petersen, R. C. (2003). Mild cognitive impairment: aging to Alzheimer's disease. Oxford; New York: Oxford University Press

[110] Breen, D. P., Michell, A. W. & Barker, R. A. (2011). Parkinson's disease –the continuing search for biomarkers. *Clin Chem Lab Med*, *49*, 393-401

[111] Ross, G. W., Abbott, R. D., Petrovitch, H., Tanner, C. M., Davis, D. G., Nelson, J., Markesbery, W. R., Hardman, J., Masaki, K., Launer, L. & White, L. R. (2006). Association of olfactory dysfunction with incidental Lewy bodies. *Mov Disord.*, *21*(12), 2062-7

[112] Lerner, A. & Bagic, A. (2008). Olfactory pathogenesis of idiopathic Parkinson disease revisited. *Mov Disord.*, *23*(8), 1076-84

[113] Del Tredici, K., Rüb, U., De Vos, R. A., Bohl, J. R. & Braak, H. (2002). Where does parkinson disease pathology begin in the brain? *J Neuropathol Experiment Neurol.*, *61*(5), 413-26

[114] Kang, J. H., Irwin, D. J., Chen-Plotkin, A. S., Siderowf, A., Caspell, C., Coffey, C. S., Waligórska, T., Taylor, P., Pan, S., Frasier, M., Marek, K., Kieburtz, K., Jennings, D., Simuni, T., Tanner, C. M., Singleton, A., Toga, A. W., Chowdhury, S., Mollenhauer, B., Trojanowski, J. Q., Shaw, L. M. & the Parkinson's Progression Markers Initiative. (2013). Association of Cerebrospinal Fluid β-Amyloid 1-42, T-tau, P-tau181, and α-Synuclein Levels With Clinical Features of Drug-Naive Patients With Early Parkinson Disease. *JAMA Neurol.*, doi, 10.1001/jamaneurol.2013.3861

[115] van Dijk, K. D., Teunissen, C. E., Drukarch, B., Jimenez, C. R., Groenewegen, H. J., Berendse, H. W. & van de Berg, W. D. (2010). Diagnostic cerebrospinal fluid biomarkers for Parkinson's disease, a pathogenetically based approach. *Neurobiol Dis.*, *39*(3), 229-41

[116] Wilhelmus, M. M. M. (2011). Apolipoprotein and LRP1 increase early in Parkinson's disease pathogenesis. *Am J Pathol*, *179*(5), 2152-2156

[117] Marin, R., Rojo, J. A., Fabelo, N., Fernandez, C. E. & Diaz, M. (2013). Lipid raft disarrangement as a result of neuropathological progresses: a novel strategy for early diagnosis? *Neuroscience.*, *245*, 26-39

[118] Hur, J. Y., Teranishi, Y., Kihara, T., Yamamoto, N. G., Inoue, M., Hosia, W., Hashimoto, M., Winblad, B., Frykman, S. & Tjernberg, L. O. (2012). Identification of novel γ-secretase-associated proteins in

detergent-resistant membranes from brain. *J Biol Chem.*, *287*(15), 11991-2005

[119] Galasko, D. & Montine, T. J. (2010). Biomarkers of oxidative damage and inflammation in Alzheimer's disease. *Biomark Med.*, *4*(1), 27-36

[120] Uehara, T., Nakamura, T., Yao, D., Shi, Z. Q., Gu, Z., Ma, Y., Masliah, E., Nomura, Y. & Lipton, S. A. (2006). S-nitrosylated protein-disulphide isomerase links protein misfolding to neurodegeneration. *Nature*, *441*(7092), 513-7

[121] Cho, D. H., Nakamura, T., Fang, J., Cieplak, P., Godzik, A., Gu, Z. & Lipton, S. A. (2009). S-nitrosylation of Drp1 mediates beta-amyloid-related mitochondrial fission and neuronal injury. *Science*, *324*(5923)., 102-5

[122] Kwak, Y. D., Wang, R., Li, J. J., Zhang, Y. W., Xu, H. & Liao, F. F. (2010). Differential regulation of BACE1 expression by oxidative and nitrosative signals. *Mol Neurodegener.*, *6*, 17-25

[123] Ortega, F., Stott, F. J., Visser, S. A. & Bendtsen, C. (2013). Interplay between alpha-, beta-, and gamma-secretases determines biphasic amyloid-beta protein level in the presence of a gamma-secretase inhibitor. *J Biol Chem.*, *288*, 785-792

[124] Swomley, A. M., Förster, S., Keeney, J. T., Triplett. J., Zhang. Z., Sultana, R. & Butterfield, D. A. (2013). Abeta, oxidative stress in Alzheimer disease: Evidence based on proteomics studies. *Biochim Biophys Acta.*, S0925-4439(13)00297-4

About the Editor

Dr. Dan Sillence
Senior Lecturer in Pharmacology
Hawthorn Building Leics LE1 9BH
United Kingdom
Tel: +44 (0)116 250 6368,
Fax: +44 (0)116 257 7287
e-mail: dsillence@dmu.ac.uk

Index

B

C

D

E

F

G

H

I

J

K

L

M

N

O

P

Q

R

T

U

V

W

X

Y